Principles of
Natural Farming

BIKASH RANJAN RAY

CONTENTS

PREFACE

Natural farming is based on the principles of natural intelligence of human beings and on the basis of traditional agricultural knowledge of the indigenous people. Natural farming (polyculture or biodiversity-based agriculture) is exactly the opposite concept of conventional agriculture (monoculture or anti-biodiversity agriculture). Natural farming is a holistic knowledge (multidisciplinary and interdisciplinary knowledge) while monoculture is based on "Cartesian Reductionism" of modern science. Agricultural graduates, who are formally trained in chemical input based monoculture, find it very difficult to understand the logic behind natural farming. In this book, an attempt has been made to present the principles of natural farming, from a holistic point of view. Holistic knowledge is a comprehensive understanding of a subject or situation, encompassing all aspects and perspectives, rather than focusing solely on individual parts or specific details. It emphasizes the interconnectedness and interdependence of various elements, leading to a more complete and nuanced view of the whole. Natural farming is based on traditional agricultural knowledge of the indigenous people, which still remains personal, oral and generational. Individual innovators tried to interpret traditional agricultural knowledge, according to their own understanding (the story of Blind Men and the Elephant), to develop new models of ecological agriculture.

Here, in this book, traditional agricultural knowledge was characterised, classified and organised in a systematic way, to derive the principles of natural farming. My hope is that this book serves as a guide for farmers, researchers, and agricultural practitioners to adopt and innovate upon the principles of natural farming. By doing so, we can collectively contribute to a resilient and sustainable agricultural future that benefits both people and the planet.

CHAPTER 1

INTRODUCTION TO NATURAL FARMING

Fundamentally, there are two types of agriculture in the world, monoculture and polyculture. Conventional agriculture (monoculture or anti-biodiversity agriculture) is based on "Cartesian Reductionism" of modern science. It divides the whole knowledge of agriculture into a number of applied science disciplines such as plant breeding (botany, genetics), soil science (chemistry), agronomy (Jack of all trades), plant pathology (microbiology), agricultural entomology (zoology), agricultural economics (economics), agricultural extension (sociology) and agricultural engineering (physics, engineering). Agricultural graduates specialised in a particular discipline tend to study, research and work in their respective discipline, ultimately for profit maximisation of the seed, input and machine manufacturing industries.

In contrast, polyculture (biodiversity-based agriculture) is based on traditional agricultural knowledge of the indigenous people. Traditional knowledge is a "Holistic Knowledge" (multi-disciplinary and inter-disciplinary knowledge). Farmers need to have knowledge in botany, zoology, microbiology, ecology, ecological economics, business management, marketing management and sociology, to succeed in modern day polyculture agribusiness.

There are innumerable types of polyculture systems in the world, which can be broadly classified into two main types, (1) Traditional agricultural landscape and (2) Modern ecological agriculture, which are developed by individual innovators, and based on traditional agricultural knowledge of the indigenous people. "Natural farming" is a system of modern ecological agriculture, developed by Masanobu Fukuoka, a Japanese farmer during the 1970s. Here, in this book, the terms "Natural Farming", "Polyculture" and

"Ecological Agriculture" were used synonymously, to represent all types of traditional agricultural landscapes and all types of modern ecological agriculture, to derive the fundamental principles of natural farming.

1.1 Definitions of Agriculture:

Agriculture is an academic subject which can be defined in many different ways:

1.1.1 Art and Science:

Agriculture is the science, art, and practice of cultivating the soil, growing crops, and raising livestock for food, fiber, fuel, and other products essential to human life. The term agriculture originates from the latin words *ager* (field) and *cultura* (cultivation).

1.1.2 Process Innovation:

Agriculture is a unique process invented by the Neolithic people, about 10,000 years ago, which involve domestication, breeding and culture of living organisms such as plants, animals and microorganisms (Agrobiodiversity), and creating an artificial (man-made) ecosystem (Agroecosystem) in the open field, for production of food for humans and for farm animals as well as production of raw materials for industries such as cotton, jute, tea, coffee, cocoa, rubber etc.

1.1.3 Management Science:

Agriculture is the applied science that integrates principles from biology, ecology, chemistry, physics, and economics to optimize the production and management of plants and animals. Agriculture is a management science (Farm Management) which deals with (a) **Agrobiodiversity Management:** Agrobiodiversity management includes domestication of plants, animals and microorganisms, breeding of plants and animals, rating and ranking of plants, animals and microorganisms and portfolio management of plants and animals, (b) **Agroecosystem Management:** Agroecosystem management includes soil health management, plant health management, animal health management, human health management, environment health management and natural control management, and (c) **Marketing Management:** Marketing management includes product quality management, customer survey, marketing strategy, price discovery, eco-agritourism management.

1.1.4 Agribusiness:

Agriculture is an economic activity that involves producing, processing, and distributing agricultural goods for profit and sustenance. Agriculture is basically a business (agribusiness) of the farmers, for earning their livelihood, and the ultimate goal of this business is maximising profit for the farmers.

1.2 Objectives of Agriculture:

The major objectives of agriculture should include:

1.2.1 To Maximise Profit for the Farmers:

Agriculture is primarily a business of the farmers and the primary objective of this business (agribusiness) is maximising profit for the farmers. Here are the components of profit maximisation for the farmers: (A) **To Minimise Risk:** Agriculture is usually practised in the open field all round the year, where it encounters mainly three types of risks, (1) climate risk (high or low temperature, high or low rainfall and high wind velocity), (2) biotic risk (incidence of pests, diseases and weeds) and (3) market risk (demand supply mismatch, price fluctuation, storage and transport damage). For example, monoculture (one-crop agriculture) bears 100% risk (all eggs in one basket) and polyculture bears a risk which is proportionate to the number of crops grown in a mixed cropping system [risk (%) = (100/number of crops) X 100]. (B) **To Minimise Cost:** Agricultural systems should try to reduce the cost of cultivation of crops without affecting the crop yield. Reduction of costs ultimately increases the profit of the farmers. (C) **To Maximise Yield:** Agricultural systems should try to increase crop yield (production of crops per unit land area), to ensure food security and nutritional security of the consumers as well as to maximise the profit of the farmer. (D) **To Maximise the Quality of Products:** Farmers should try to maximise the quality of the agricultural products ultimately to maximise the price of the products and to maximise the profit of the farmers. (E) **To Maximise Product Price:** Farmers should adopt various agricultural practices which increase the quality of the produce and should follow various marketing strategies to maximise the price of their agricultural products. (F) **To Establish Multiple Sources of Income for the Farmer:** Farmers should establish multiple sources of income from their farms to minimise the risk and to maximise the profit of the farmers. (G) **To Earn Regular Passive Income:** Farmers should try to ensure a regular source of passive income, on a daily basis, from agriculture and allied activities, to maintain their livelihood and standard of life.

1.2.2 To Improve Human Health and Immunity:

The ultimate aim of agriculture is to provide healthy and nutritious food to every human being in the world because food is a fundamental right to all human beings. Farmers should also increase the medicinal values of food to improve public health.

1.2.3 To Improve Environment Health:

Agriculture should also contribute towards adaptation and mitigation of climate change, biodiversity loss and pollution (the triple planetary crisis) through the adoption of sustainable and resilient agricultural practices.

1.3 Types of Agriculture:

Fundamentally, there are two types of agriculture in the world, monoculture and polyculture.

1.3.1 Monoculture:

Monoculture is a system of agriculture which grows only one single crop plant species or variety in a field and tries to eliminate all other plants (weeds), animals (pests) and microorganisms (pathogens) from that field, using deep tillage and applying toxic chemical herbicides, pesticides, fungicides and antibiotics, to maximise yield.

Characteristics of Monoculture:

The Anthropocene: The Anthropocene is a proposed geological epoch characterized by significant human impact on Earth's geology and ecosystems. Coined by Paul Crutzen in 2000, it suggests that human activities, particularly since the Industrial Revolution, have drastically altered the planet's climate, biodiversity, and natural processes. Key markers include accelerated climate change, mass extinction rates exceeding historical norms, and widespread pollution.

Cartesian Reductionism: Cartesian reductionism is a philosophical approach derived from the ideas of René Descartes, a 17th-century French philosopher and mathematician. It emphasizes breaking down complex phenomena into smaller, simpler components to understand the whole. This method has significantly influenced the development of science, mathematics, and philosophy. Scientists broke down traditional holistic knowledge of agriculture into a number of various disciples of applied science such as plant breeding (botany, genetics), soil science (chemistry), agronomy (Jack of all trades), plant pathology (microbiology), agricultural entomology (zoology), agricultural economics (economics), agricultural extension (sociology), and agricultural engineering (physics, engineering).

Anti–Biodiversity Agriculture: Monoculture intends to eliminate the biodiversity of organisms such as plants (weeds), animals (pests, natural enemies and earthworms) and microorganisms (plant pathogens, antagonists, arbuscular mycorrhizal fungi and plant growth promoting rhizobacteria) from agricultural fields, by applying deep tillage and toxic chemicals, to maximise yield and to maximise profit of the input manufacturing industries.

Completely Disturbed Ecosystem: Monoculture intentionally creates a completely disturbed ecosystem and a vacuum of biodiversity within the biosphere. The aim is to eliminate all ecosystem services from the agroecosystem, to maximise the dependence on synthetic chemical inputs. However, a vacuum of biodiversity is not sustainable in nature, therefore invasion of biodiversity (in the form of weeds, pests and pathogens) is inevitable in monoculture.

Input-Based Agriculture: The ultimate objective of monoculture is to maximise the use of inputs (commercial seed, fertilisers, micronutrients, plant growth regulators, herbicides, pesticides, fungicides and antibiotics) in agriculture, for maximising yield and for maximising profit of the seed, input and machine manufacturing industries. Monoculture intends to replace traditional knowledge based agriculture with input based agriculture.

Industrial Agriculture: Monoculture is often referred to as industrial agriculture because monoculture supports mechanisation of agriculture and industrialisation of agriculture (invasion of industry in the domain of agriculture). Monoculture thus helped create a monopoly of industries in the global agrifood business.

Unsustainable Agriculture: Monoculture is unsustainable agriculture because it has serious detrimental impacts on the environment, soil health, biodiversity, and long-term productivity and human health: (1) **Soil Degradation**: Continuous monoculture depletes specific nutrients from the soil, leading to nutrient imbalances and reduced fertility. Heavy use of synthetic fertilizers to compensate for nutrient loss further degrades soil structure and microbial life. In fact, 30% agricultural land has already been desertified by monoculture, soil erosion and climate change and if this trend continues, 90% land can be desertified in 2050 (FAO). (2) **Pest and Disease Vulnerability**: Large-scale monoculture creates ideal conditions for pests and diseases to spread quickly. This necessitates heavy pesticide use, which harms ecosystems and fosters pesticide-resistant pests. (3) **Loss of Biodiversity**: Monoculture replaces diverse ecosystems with uniform crop fields, reducing plant, animal, and microbial diversity. This loss of biodiversity weakens ecosystems, making them less resilient to climate change and other disturbances. (4) **Dependence on Chemical Inputs**: Monoculture farming often relies on synthetic inputs like fertilizers, herbicides, and pesticides to maintain productivity. These chemicals pollute water sources, harm non-target species, and contribute to soil and air pollution. (5) **Water Resource Depletion**: Monoculture crops often require intensive irrigation, leading to the over-extraction of groundwater and depletion of water resources. (6) **Climate Change Contribution**: Practices like deforestation for monoculture and soil degradation release significant amounts of carbon dioxide and other greenhouse gases. (7) **Economic Risks**: Farmers growing a single crop are vulnerable to market fluctuations, pests, and extreme weather, which can lead to financial instability.

Objectives of Monoculture:

The objectives of monoculture include:

Monopoly of Input Based Agriculture:

The primary objective of monoculture is to maximise the use of chemical fertilisers, chemical pesticides and commercial hybrid seeds, to maximise the profit of the input manufacturing industries. Ultimately to establish a monopoly of monoculture, all over the world.

Industrialization of Agriculture:

The industrialization of agriculture refers to the transformation of farming from traditional methods into a modern, mechanized, and large-scale industry. This process involves the integration of advanced technologies, capital-intensive methods, and corporate strategies to enhance productivity and efficiency in agriculture. Here are some key aspects of agricultural industrialization: (1) **Mechanization:** Use of machinery like tractors, harvesters, and drones to reduce labor and increase efficiency. Automated irrigation systems for precise water management. (2) **Chemical Inputs:** Extensive use of synthetic fertilizers, pesticides, and herbicides to boost crop yields. Concerns about soil health and environmental sustainability. (3) **High-Yield Varieties (HYVs):** Development and use of genetically modified (GM) seeds and HYVs for better crop output. Focus on monoculture to maximize productivity of selected crops. (4) **Large-Scale Production:** Consolidation of small farms into large agribusinesses. Specialization in specific crops or livestock to achieve economies of scale. (5) **Global Supply Chains:** Integration with global markets, resulting in export-oriented farming. Dependency on international trade policies and market dynamics. (6) **Technology Integration:** Use of precision farming, satellite imaging, and IoT for real-time monitoring of farm activities.

Data analytics for crop forecasting and resource optimization. (7) **Corporate Involvement:** Shift from family-owned farms to corporate farming. Investment by agribusiness companies in the entire value chain, including production, processing, and distribution.

To Leverage Land and Labour of the Farmers:

The business plan of the chemical industries was to leverage the land and labour of the farmers, to maximise the use of synthetic inputs in agriculture, to maximise their profit. The plan includes: (1) Persuading the farmers to leave their own country plough and to depend totally on hired tractors for tilling their land. This helps profit maximisation of the tractor manufacturing industries and tractor hiring service providers, (2) Deep tillage with tractors destroys soil horizons, soil structure, soil organic matter, water holding capacity of soil, soil biodiversity, soil health and ecosystem services of soil, which makes the farmers completely dependent on chemical fertilisers, for crop production, that help profit maximisation of the fertiliser manufacturing industries. (3) Tillage destroys water holding capacity of soil, which maximises the profit of the irrigation equipment manufacturing industries, (4) Application of chemical fertilisers and application of irrigation increased incidence of insect pests, diseases and weeds, application of pesticides eliminate the natural enemies (predators and parasitoids) and antagonists from agricultural fields and pests develop resistance against pesticides, fungicides, antibiotics and herbicides. This situation helps profit maximisation of the pesticides manufacturing industries. (5) Farmers and their family members work in the field, free of cost, for carrying out monoculture for the benefit of consumers and industries. Industries or governments do not pay any salary or wages to the farmers for their noble service.

To Leverage Capital of the Governments:

As a clever business plan, industries leverage the capital of the governments, to maximise their profit from monoculture. (1) Industries do not incur any cost for promotion of monoculture. Governments bear all the costs of education, research, extension and administration of monoculture. (2) Governments provide direct and indirect subsidies to the farmers and industries for providing free electricity, free irrigation, subsidised tractors, machines and tools, seeds, inputs, storage, transport, regulated market, public distribution system (PDS) and minimum support price (MSP). (3) Governments also provide cash incentives directly to the bank accounts of the farmers, to compensate for their losses in practicing monoculture. (4) Governments also pay crop insurance premiums and pay compensations to the farmers for crop losses due to climate change and extreme weather events.

Food Security of Consumers:

Government's objective of adopting monoculture as their official form of agriculture and neglecting the traditional system of agriculture, was to ensure food security of their growing population of consumers (citizens) at an affordable (artificially low, market distorting) price, by providing subsidies from public funds. However, food grain production has already stagnated or is declining (The Law of Diminishing Marginal Returns). And 30% agricultural land in the world has already been desertified by monoculture and climate change, threatening future food security.

Materials and Methods of Monoculture:

Industries and governments engaged non-farmer urban scientists since the 1960s, to develop the materials and methods of monoculture, for profit maximisation of the seed, input and machine manufacturing industries. Here are the main materials and methods used in monoculture:

Mechanised Tillage: Machine manufacturing industries promote tractors and other accessories for deep tillage of agricultural land, for clean cultivation. Mechanised tillage has several advantages and disadvantages. Mechanised tillage saves time, cost and labour requirement. Deep tillage, on the other hand, destroys soil horizons, soil structure (soil aggregates), soil organic matter content, water holding capacity of soil, soil ecosystem, soil biodiversity, soil health and promotes soil erosion, soil degradation, desertification of land and soil compaction. Tillage also helps build parasite seed banks in soil and help dispersal of parasitic plant seeds along with tilling blades. Mechanised tillage needs burning of fossil fuel and emission of greenhouse gases and increases cost of cultivation of crops.

HYV, Hybrid or GM Seed: Seed companies promote commercial seeds of high yielding varieties (HYV), hybrid seeds (F1) or genetically modified (GM) seeds, and discourage use of traditional seeds, to the farmers, to increase yield. Hybrid seeds lose their hybrid vigour in the next (F2) generation, therefore farmers cannot retain hybrid seeds for sowing in the next cropping season. Each year, farmers have to buy costly commercial seeds from the seed companies to continue monoculture. In this way, seed companies create a monopoly to maximise their profit. The objective of breeding these hybrid seeds usually yields higher yields, therefore, these crops are often poor in taste, nutritive value and medicinal value. Low genetic bases of these hybrid seeds help genetic erosion, which threaten future food security.

Irrigation: Irrigation is an essential requirement in monoculture. Deep tillage reduces the water holding capacity of soil. And a huge amount of water is required to dissolve the chemical fertilisers applied in soil, to make them available to plants. Usually, water is sourced from underground aquifers to provide irrigation water to the crop plants. In fact, about 70% of the total water lifted from the underground sources in the world is used in agriculture and allied activities which depletes underground water and threatens supply of drinking water to the citizens. Irrigation with underground water increases salinity in soil, increases arsenic (heavy metal), chlorine and other harmful content in water that can potentially cause health hazards. Monoculture therefore is helpful for profit maximisation of the irrigation equipment manufacturing industries.

Chemical Fertilisers: Monoculture is absolutely dependent on synthetic chemical fertilisers and chemical micronutrients. Monoculture recommends deep tillage with tractors for clean cultivation, which completely eliminate plant biodiversity (weeds), animal biodiversity (pests, earthworms) and microorganism biodiversity (pathogens, antagonists, arbuscular mycorrhizal fungi and plant growth promoting rhizobacteria) from agricultural fields. In absence of soil biodiversity and ecosystem services the farmers have to completely depend on synthetic chemical fertilisers, micronutrients and synthetic plant growth regulators for achieving crop growth and development. Chemical fertilisers change chemical properties of the soil, change pH of the soil, cause emission of greenhouse gases (NO_2) from soil, leaching of chemicals to underground water and eutrophication of water bodies. Luxury consumption of chemical fertilisers cause unusual crop growth that helps severe incidence of insect pests, diseases and weeds.

Micronutrients: Micronutrients are the chemical elements and compounds which are required by the plants in very small quantities for healthy growth and development. Micronutrients include boron (B), chlorine (Cl), copper (Cu), iron (Fe), manganese (Mn), molybdenum (Mo), zinc (Zn), and nickel (Ni). Monoculture uses of high doses of nitrogen (N), phosphorus (P) and potassium (K) fertilisers in soil which results in higher uptake of micronutrients from soil by plants (luxury consumption). This luxury consumption of chemical fertilisers by plants causes micronutrient deficiency in some soils and sometimes application of overdoses of micronutrients in soil causes excess situations. Both deficiency and excess of micronutrients in soil causes typical symptoms in crop plants. This man–made micronutrient imbalance in soil helps profit maximisation of the chemical industries.

Plant Growth Regulators: In addition to fertilisers and micronutrients, monoculture needs synthetic plant growth regulators to boost plant growth. Under natural soil conditions, plant growth promoting rhizobacteria (PGPR) provide required growth regulators to the crop plants for optimum plant growth and development. Monoculture applies deep tillage to destroy soil biodiversity and ecosystem services. Therefore, synthetic plant growth regulators become essential in monoculture to boost crop growth. This situation helps profit maximisation of the chemical industries.

Herbicides: Herbicide became an essential input in monoculture. Because monoculture provides an ideal environment for luxurious growth and survival of weed biodiversity. Monoculture provides a wide biodiversity-free gaps in between two rows of crop plants for competition-free growth of the weeds. Monoculture provides abundant supply of chemical fertilisers and irrigation water to the weeds for uncontrolled growth. Chemical herbicides are the only low–cost solution available to the farmers, for weed control in monoculture. Herbicides are synthetic chemical compounds which can selectively kill weeds, leaving the crop plants unharmed. Herbicide-resistant crop plant varieties have also been developed to maximise the use of herbicides. However, weeds quickly develop resistance against herbicides through the process of natural selection and become uncontrollable with single herbicides. This situation favours profit maximisation of the chemical (herbicide manufacturing) industries.

Pesticides: Pesticides (insecticides, acaricides, nematicides, rodenticides) are essential inputs in monoculture. Pesticides are synthetic chemical compounds which are intended to kill insect pests of crop plants. However, by default, they also kill the natural enemies (predators and parasitoids of pests) and beneficial insects (such as honey bees and pollinators) in the agricultural fields. Elimination of natural enemies from the environment helps exponential growth of pests and development of pest epidemics on crops in monoculture. Susceptibility of crop plants further increases the requirement of pesticides in monoculture. Insect pests quickly develop resistance against the pesticides through natural selection that further help maximise the use of insecticides. Farmers neither need any licence from the government to handle highly toxic chemical insecticides (poisons) nor they need any prescription from the registered agricultural practitioner (agriculture graduate) to use the insecticides at the right dose. In this way pesticides help profit maximisation of the chemical (pesticide manufacturing) industries.

Fungicides: Fungicides are essential inputs in monoculture. Fungicides are synthetic chemical compounds which can kill plant pathogenic fungi as well as their antagonistic fungi and beneficial fungi species. Monoculture provides a congenial environment for incidence and spread of pathogenic fungi and diseases in crops. Susceptible crop plants, high density planting, high doses of chemical fertilisers and irrigation water help rapid spread of fungal diseases in monoculture. Chemical fungicides are the only solutions available to the farmers to save the crops from fungal diseases in monoculture. Fungal pathogens however quickly develop resistance against the fungicides through natural selection and farmers need newer fungicides or combinations of fungicides, at regular intervals, to control plant pathogens. This situation helps profit maximisation of the chemical (fungicide manufacturing) industries.

Antibiotics: Synthetic antibiotics have now become a common input in monoculture. Antibiotics are synthetic chemical compounds which can kill plant pathogenic bacteria as well as other antagonistic or beneficial bacteria in agricultural fields. Monoculture provides an ideal environment for rapid spread of bacterial diseases. Susceptible crop plant variety, high density planting, high doses of chemical fertilisers and irrigation water help rapid spread of bacterial diseases in monoculture. Antibiotics are the only solutions available to the farmers, to control bacterial diseases in monoculture crops. Plant pathogenic bacteria quickly develop resistance against antibiotics through natural selection. These antibiotic-resistant bacteria are often called superbugs, which are a threat the public health. Synthetic antibiotics help profit maximisation of chemical (antibiotic manufacturing) industries.

Crop Insurance: Crop insurance is essential in monoculture. Monoculture systems (all eggs in one basket) bears the highest (100%) possible risk of crop loss, due to climate change, extreme weather events and due to occurrence of insect pests, diseases and weeds. Therefore, monoculture helps profit maximisation of the crop insurance companies.

Mechanised Harvesting: Monoculture crops are suitable for mechanised harvesting, threshing, winnowing and bagging. Mechanised harvesting is beneficial for the farmers due to saving of time and saving labour cost. Mechanised harvesting helps profit maximisation of the machine manufacturing industries and the machine hiring service providers.

Transportation: Monoculture crops need to be transported to a long distance for storing or marketing. This situation helps profit maximisation of the transport service providers.

Storage and Cold Storage: Agricultural products are usually highly perishable commodities. It needs storing or cold storing for a long time to get a better market price. This situation help profit maximisation of the storage or cold storage service providers.

Centralised Marketing: Monoculture products are produced in bulk which are beyond the capacity of the farmers to market and they needs specialised third party marketing facilities. These entities are called marketing intermediaries or middlemen. Often farmers get a small fraction of the consumer price of their products and these middlemen capture the lion's share of the consumer price of agricultural commodities. This centralised marketing system helps profit maximisation of the marketing intermediaries and the food processing industries.

Credit: Farmers need funds on credit to implement monoculture, because the cost of cultivation of crops is very high in monoculture. The credit worthiness of small-scale farmers are very low so they need to borrow money from the village money lenders at an exorbitant rate of interest (from 60 to 120% per year). Monoculture crops are harvested at a time in the season creating an oversupply situation (glut) in the market. Farmers remain in a hurry to repay their debts and they are forced to sell the crops at a low price. Total crop loss is common in monoculture due to extreme weather events like drought, excess rainfall or storm. Small-scale farmers therefore get trapped in a vicious cycle of debt trap. This situation help profit maximisation of village money lenders.

Trade-offs in Monoculture:

Monoculture is a trade-off between food security and environmental disaster. Here are some harmful effects of monoculture:

Deforestation: Monoculture farming significantly contributes to deforestation. (1) **Land Clearing for Large-Scale Farming**: To establish vast monoculture plantations, forests are often cleared, removing native vegetation and destroying habitats. Examples include deforestation in the Amazon rainforest for soybean cultivation and oil palm plantations in Southeast Asia. (2) **Loss of Biodiversity**: Forests are biodiversity hotspots, supporting a wide variety of plant and animal species. When forests are cleared for monoculture, this biodiversity is lost. (3) **Increased Demand for Arable Land**: Monoculture depletes soil nutrients over time, making the land less fertile and prompting farmers to clear additional forest areas for cultivation. (4) **Commercial Expansion**: Rising global demand for cash crops like palm oil, coffee, sugarcane, and rubber drives deforestation to expand monoculture plantations. (5) **Soil Degradation and Erosion**: Monoculture weakens the soil structure and accelerates erosion, making the land unsuitable for long-term farming. Farmers then clear forests to access fertile land. (6) **Water Resource Stress**: Monoculture crops, especially water-intensive ones like sugarcane and cotton, deplete local water sources. This drives deforestation as farmers seek new water–rich areas.

Genetic Erosion: Monoculture farming significantly contributes to genetic erosion, which refers to the loss of genetic diversity within plant and animal species. In monoculture systems, farmers focus on growing a single crop variety, often genetically identical, leading to a dramatic reduction in genetic variation. This practice has serious implications for agricultural resilience and long-term sustainability. How Monoculture Causes Genetic Erosion: (1) **Genetic Homogeneity**: In monoculture, a single crop variety is planted across large areas, leading to genetic uniformity. If a single crop variety is used over successive planting seasons, any genetic diversity within that crop is reduced, making it more susceptible to pests, diseases, and changing environmental conditions. (2) **Loss of Traditional Varieties**: Monoculture often pushes out traditional, local, and heirloom crop varieties, which were bred to be resilient to local pests, diseases, and climate variations. These varieties contain unique genetic traits that are vital for future breeding and adaptation to changing environmental conditions. (3) **Increased Vulnerability to Diseases and Pests**: Lack of genetic diversity means a single disease or pest outbreak can wipe out an entire crop. For example, the Irish Potato Famine in the 19th century was caused by a single crop variety's vulnerability to a potato blight. (4) **Reliance on Commercial Seed Varieties**: Many monoculture farms rely on commercial, genetically modified, or hybrid seeds that are often patented by large corporations. This leads to a reduction in the variety of seeds available to farmers and a decrease in the genetic pool for future breeding. (5) **Narrow Genetic Base in Breeding Programs**: To ensure high yields, modern breeding programs often focus on a small number of traits (e.g., high yield, pest resistance). This narrow focus reduces the genetic diversity of crop varieties used in monoculture farming and limits the potential for future improvements.

Soil erosion: Soil erosion due to monoculture is a significant environmental concern. Monoculture, the practice of growing a single crop over a large area, leads to the degradation of soil structure and increases vulnerability to erosion by wind and water. This practice strips the land of its protective cover and reduces biodiversity, causing long-term harm to agricultural productivity and ecosystems. How Monoculture Causes Soil Erosion: (1) **Loss of Vegetative Cover**: After harvest, monoculture fields are often left bare, exposing soil to wind and rain. Absence of diverse root systems weakens soil structure. (2) **Tillage**: Monoculture follows seep tillage with tractors for clean cultivation. Deep tillage destroys soil horizons, soil structure (soil aggregates), organic matter content in soil, and water holding capacity of soil. This loosening of soil leads to soil erosion. (3) **Soil Nutrient Depletion**: Continuous cultivation of the same crop depletes specific nutrients, reducing soil fertility. (4) **Reduced Organic Matter**: Lack of crop rotation minimizes the addition of organic matter, leading to poor soil health. Without organic matter, the soil cannot retain moisture, increasing susceptibility to erosion. (5) **Compacted Soil:** Repeated use of heavy machinery in monoculture compresses the soil, reducing water infiltration and increasing surface runoff. (6) **Weed and Pest Management:** Monoculture often relies heavily on chemical herbicides, which can kill beneficial ground-cover plants and microorganisms, leaving soil bare and less resilient.

Soil degradation: Monoculture, the practice of growing a single crop over a large area continuously, leads to significant soil degradation. Monoculture disrupts natural soil processes, reduces fertility, and depletes essential nutrients, ultimately jeopardizing long-term productivity and environmental sustainability. How Monoculture Causes Soil Degradation: (1) **Nutrient Depletion:** Repeated cultivation of the same crop removes specific nutrients from the soil. The absence of crop rotation prevents nutrient replenishment, leading to imbalances. (2) **Loss of Soil Biodiversity:** Monoculture discourages diverse microorganisms and beneficial organisms like earthworms, crucial for healthy soil. Reduced microbial diversity affects soil structure and fertility. (3) **Reduced Organic Matter:** Continuous cropping and removal of plant residues lower organic matter content. This diminishes the soil's ability to retain water and nutrients. (4) **Soil Erosion:** Monoculture fields are often left bare after harvest, exposing soil to wind and water erosion. Lack of crop diversity weakens root systems, making soil more prone to displacement. (5) **Increased Soil Compaction:** Repeated use of heavy machinery in monoculture farms compresses soil layers. Compacted soil reduces water infiltration, aeration, and root growth. (6) **Dependency on Chemicals:** Monoculture often requires synthetic fertilizers and pesticides to maintain yields. Over time, chemical residues degrade soil quality and kill beneficial organisms. (7) **Loss of Soil Structure:** Monoculture disrupts natural soil aggregation, leading to crusting and hardpan formation. Poor soil structure results in reduced water retention and drainage.

Desertification of land: Monoculture farming significantly contributes to desertification, a process where fertile land becomes arid and unproductive, often turning into desert-like conditions. This environmental degradation is primarily driven by poor land management practices and the ecological impacts of monoculture farming, including soil erosion, nutrient depletion, and water scarcity. According to the FAO of the UN, 30% of agricultural land in the world has already been desertified, and if this trend continues, 90% land can be desertified in 2050.

Depletion of underground water: The depletion of underground water by agriculture is a critical global issue, driven by over-extraction of groundwater to meet irrigation demands. Agriculture accounts for about 70% of global freshwater withdrawals, with much of it sourced from underground aquifers. This overuse has severe environmental, economic, and social consequences. Causes of Groundwater Depletion in Agriculture: (1) **Inefficient Irrigation Practices:** Over-irrigation and lack of water-saving techniques lead to excessive groundwater use. Methods like flood irrigation waste significant amounts of water. (2) **High Water-Intensive Crops:** Crops like rice, sugarcane, and cotton require large quantities of water. Cultivation of such crops in water-scarce regions exacerbates the problem. (3) **Monsoon Dependency:** Inconsistent rainfall forces farmers to rely on groundwater for irrigation. (4) **Climate Change:** Rising temperatures and unpredictable weather patterns increase water demand for crops. (5) **Policy Gaps and Subsidies:** Subsidies for electricity used in pumping groundwater encourage over-extraction. Lack of regulation on groundwater extraction leads to unregulated use.

Carbon Footprint: A carbon footprint in agriculture refers to the total greenhouse gas (GHG) emissions produced throughout the agricultural lifecycle. These emissions primarily include carbon dioxide (CO_2), methane (CH_4), and nitrous oxide (N_2O), which contribute to global warming. Understanding and managing the carbon footprint in agriculture is crucial for promoting sustainable farming practices. Major Contributors to Carbon Footprint in Agriculture: (1) **Soil Management**: Overuse of synthetic fertilizers emits N_2O, a potent GHG. Excessive tilling releases stored CO_2 from the soil. (2) **Livestock**: Enteric fermentation in ruminants (like cows) emits CH_4. Manure decomposition releases CH_4 and N_2O. (3) **Energy Use**: Fossil fuels used for machinery, irrigation pumps, and transportation emit CO_2. (4) **Land Use Changes**: Deforestation for agricultural expansion leads to loss of carbon sinks. Peatland conversion releases significant amounts of CO_2. (5) **Crop Production**: Paddy fields generate CH_4 due to anaerobic decomposition in flooded conditions. Burning of crop residues releases CO_2 and particulate matter. (6) **Post-Harvest Activities**: Energy-intensive processes like storage, processing, and transportation add to the footprint.

Climate change: Monoculture farming is a significant contributor to climate change due to its environmental impact, particularly in terms of carbon emissions, deforestation, soil degradation, and resource-intensive practices. (1) **Carbon Emissions from Soil Degradation**: (a) **Soil Erosion and Depletion**: Monoculture farming, especially when it involves the continuous cultivation of the same crop, depletes the soil of essential nutrients, organic matter, and microorganisms. As soil quality deteriorates, it becomes less effective at storing carbon, and in some cases, the carbon stored in the soil is released back into the atmosphere as carbon dioxide (CO_2), contributing to global warming. (b) **Loss of Soil Carbon Sequestration**: Healthy soils act as carbon sinks, absorbing and storing carbon from the atmosphere. Monoculture practices that degrade soil structure and reduce organic matter decrease the soil's ability to sequester carbon, thus releasing more CO_2 into the air. (2) **Deforestation for Agricultural Expansion**: To make way for large monoculture farms (such as palm oil, soy, or corn plantations), forests and natural ecosystems are often cleared. This deforestation leads to a loss of carbon sinks because trees absorb large amounts of CO_2. Additionally, the clearing process releases stored carbon from trees and vegetation into the atmosphere, accelerating climate change. (3) **Use of Synthetic Fertilizers and Pesticides:** Monoculture farming systems rely heavily on chemical inputs, such as synthetic fertilizers and pesticides. The production and application of fertilizers, particularly nitrogen-based fertilizers, release nitrous oxide (N_2O), a potent greenhouse gas. Furthermore, the use of chemicals contributes to soil and water pollution, which can degrade local ecosystems and exacerbate environmental stressors. (4) **Increased Energy Use:** Monoculture farms typically require high energy input for machinery, irrigation, transportation, and chemical applications. The reliance on fossil fuels for these activities leads to higher emissions of greenhouse gases, further contributing to climate change. (5) **Loss of Biodiversity and Ecosystem Services:** Monoculture systems support little to no biodiversity, which reduces the natural ecosystem services that mitigate climate change. Healthy ecosystems (such as wetlands, forests, and diverse agricultural landscapes) provide essential services like carbon sequestration, water regulation, and temperature moderation. The loss of these services can amplify the effects of climate change. (6) **Water Use and Water Scarcity:** Monoculture farming often relies on intensive irrigation, leading to the depletion of local water resources. The over-extraction of water, especially in arid regions, can reduce the land's capacity to support agriculture in the long term and increase the vulnerability of ecosystems to climate-related stresses, such as droughts.

Biodiversity loss: Monoculture is a major contributor to biodiversity loss. By prioritizing uniformity and high yields, monoculture disrupts natural ecosystems, diminishes habitat availability, and undermines the complex interactions that sustain life. How Monoculture Causes Biodiversity Loss: (1) **Habitat Destruction:** Large-scale monoculture often involves clearing forests, grasslands, and wetlands, destroying habitats for countless species. It replaces diverse ecosystems with uniform crop fields, leaving little room for native flora and fauna. (2) **Reduction of Genetic Diversity:** Monoculture relies on a few high-yield crop varieties, leading to a significant reduction in genetic diversity within agricultural systems. Low genetic diversity makes crops and ecosystems more vulnerable to diseases, pests, and climate change. (3) **Pesticide and Herbicide Use:** The extensive use of chemicals to manage pests and weeds in monoculture kills non-target species, including beneficial insects, birds, and soil microorganisms. Pollinator populations, such as bees and butterflies, are particularly affected. (4) **Disruption of Ecosystem Services:** Monoculture weakens natural services like pollination, pest control, and nutrient cycling by eliminating the biodiversity that supports these functions. (5) **Soil Microbial Imbalance:** Monoculture depletes soil biodiversity by disrupting microbial communities that contribute to soil health and plant growth. (6) **Water Pollution:** Chemical runoff from monoculture farms contaminates water bodies, harming aquatic ecosystems and reducing biodiversity in rivers, lakes, and wetlands. (7) **Climate Change Feedback:** Monoculture contributes to deforestation and soil degradation, releasing stored carbon and exacerbating climate change, which further threatens biodiversity.

Environmental Pollution: Monoculture farming contributes significantly to environmental pollution, largely due to its heavy reliance on synthetic chemicals, resource-intensive practices, and ecological disruptions. (1) **Chemical Pollution:** (a) **Pesticides and Herbicides:** Monoculture systems are highly susceptible to pests and weeds, leading to excessive use of pesticides and herbicides. These chemicals contaminate soil, water, and air, harming non-target organisms and ecosystems. (b) **Fertilizer Runoff:** Overuse of synthetic fertilizers results in nutrient runoff, which pollutes rivers, lakes, and oceans. This causes eutrophication, leading to algae blooms and "dead zones" where aquatic life cannot survive. (2) **Soil Degradation:** Continuous monoculture depletes soil nutrients, reducing its ability to retain water and support diverse microbial life. Degraded soil often requires more fertilizers, perpetuating the pollution cycle. (3) **Water Contamination:** Fertilizers, pesticides, and herbicides leach into groundwater, contaminating drinking water sources with nitrates and harmful chemicals. Polluted water bodies harm aquatic ecosystems and human communities. (4) **Air Pollution:** (a) **Agrochemical Drift:** Pesticides sprayed on monoculture crops can drift into the atmosphere, exposing nearby communities and ecosystems to toxins. (b) **Greenhouse Gas Emissions:** Synthetic fertilizer production and application release nitrous oxide, a potent greenhouse gas. Deforestation for monoculture farming contributes to carbon dioxide emissions. (5) **Plastic Pollution:** Monoculture systems often use plastic mulch for weed control and irrigation efficiency, contributing to plastic waste in agricultural landscapes. (6) **Waste Generation:** Monoculture farming produces large quantities of crop residue, which is often burned, releasing pollutants like carbon monoxide, particulate matter, and other toxic gases.

Bioaccumulation and Biomagnification of microplastics and pesticides: Bioaccumulation and biomagnification are processes by which harmful substances, such as microplastics and pesticides, accumulate in the tissues of organisms through the food chain. Monoculture farming, with its heavy reliance on chemical pesticides and plastic use, significantly contributes to these phenomena, leading to environmental and health risks. (1) **Pesticide Use in Monoculture:** (a) **Persistent Pesticides:** Monoculture systems rely heavily on synthetic pesticides to protect crops from pests.

These pesticides often contain persistent chemicals, meaning they do not break down easily in the environment and can remain in the soil, water, and food chain for long periods. These chemicals are absorbed by plants, and when animals feed on these plants or drink contaminated water, they accumulate pesticides in their bodies. Over time, higher levels of pesticides can bioaccumulate in organisms at higher trophic levels. (b) **Runoff into Water Bodies**: Rain and irrigation runoff from monoculture farms carry pesticide residues into nearby water bodies, contaminating aquatic ecosystems. Aquatic organisms such as fish and insects ingest these pesticides from contaminated water, leading to bioaccumulation. These chemicals then biomagnify up the food chain as predators (birds, larger fish, or mammals) consume contaminated prey. (2) **Microplastics from Plastic Mulch and Agricultural Waste**: (a) **Plastic Mulch and Plastic Debris**: Monoculture farming often uses plastic mulch to retain moisture and control weeds. Over time, the plastic breaks down into microplastics, which can enter the soil and water systems. These microplastics can be ingested by soil organisms, such as earthworms, or aquatic organisms that come into contact with contaminated water or soil. (b) **Bioaccumulation of Microplastics**: Organisms that consume contaminated soil or water ingest microplastics, which accumulate in their bodies. As smaller organisms are consumed by larger predators, the microplastics biomagnify, leading to increasingly high concentrations in higher-level predators. (3) **Loss of Biodiversity in Monoculture Systems**: (a) **Reduction in Natural Predators**: Monoculture systems lack biodiversity, which means fewer natural predators are present to control pests. This results in an overreliance on chemical pesticides. As a consequence, pesticide contamination is more widespread and persistent, affecting not only target species but also non-target organisms that could have naturally regulated pest populations. (b) **Trophic Cascade**: The reduction in biodiversity leads to a disruption in the food chain. Pesticide and microplastic contamination at lower trophic levels can have cascading effects, impacting species at higher levels, leading to biomagnification of these pollutants.

Insect Pests Outbreak: Monoculture cropping systems provide abundant supply of food and habitat for specific species of herbivorous insects (pests), that help exponential growth of their populations. Chemical pesticides are applied in the field to control these pests which results in development of resistance among pests against pesticides, through natural selection and on the other hand, complete elimination of all natural enemies (predators and parasitoids) of these pests from the environment that further help their exponential growth (epidemics).

Plant Disease Outbreaks: Monoculture crop plants are inherently susceptible to local diseases due to centralised plant breeding and narrow genetic base. Excessive use of chemical fertilisers, micronutrients, plant growth regulators and irrigation in monoculture make these crop plants more succulent and less resistant to pests and diseases. High density monoculture crops provide abundant substrate and favourable growing conditions for pathogenic fungi and bacteria and ideal conditions for abundance of virus vectors. Usually chemical fungicides and synthetic antibiotics are applied on crop or in soil to control diseases. However, fungal and bacterial pathogens quickly develop resistance against fungicides and antibiotics respectively. On the other hand, these fungicides and antibiotics eliminate antagonistic fungi and bacteria (natural enemies) from the environment which help exponential growth of plant pathogens and plant diseases.

Weed Infestation: Monoculture systems were designed to maximise the growth of weed biomass in the agricultural fields and to maximise the use of chemical herbicides for profit maximisation of the herbicide manufacturing industries. Large inter-plant vacant places in monoculture encourage profuse growth of unwanted plant species (weeds) in agricultural fields. Tillage helps preservation of weed seeds underground and staggered germination of the seed with every tillage operation. Chemical fertilisers and irrigation water help abundant growth of weeds in monoculture crop fields. Chemical herbicides are commonly used to control weeds in monoculture fields. However, herbicides quickly develop herbicide-resistant weeds (superweeds) which cannot be controlled with any herbicide.

Pollinator Decline: Pollinator populations are experiencing significant declines globally, impacting biodiversity and agricultural productivity. Studies indicate that bee populations in some regions have decreased by over 60% in 15 years, with specific species like the carpenter bee declining by 97%. In Britain, approximately one-third of wild pollinator species have shown declines since 1980. Major threats include habitat loss, pesticide exposure, and climate change, which collectively threaten food security and ecosystem health (Powney GD et al. 2019).

High Cost of Cultivation of Crops: The cost of cultivation of crops in monoculture is extremely high, usually therefore the cost of cultivation includes only the input costs and the costs of hired labour, to deceive the innocent farmers. The costs that are not usually included in the cost of cultivation of crops include: (1) Farmers salary, (2) Wages of family labour, (3) Annual rent of land, (4) Interest paid on borrowed capital, (5) Environmental cost of production of crops, and (6) Subsidies provided by the government in education, research, extension and administration of monoculture, indirect subsidy in electricity, irrigation, tractor, equipments, seed, fertilisers, pesticides, transport, storage, marketing, minimum support price (MSP), crop insurance, and payment of direct cash incentives to the farmers.

Human Diseases: Monoculture contribute to the spread and development of human diseases by creating imbalanced ecosystems, increasing chemical use, and reducing dietary diversity. How Monoculture Contributes to Human Diseases: (1) **Increased Pesticide and Herbicide Use:** Monoculture crops are more vulnerable to pests and diseases, leading to heavy reliance on chemical pesticides and herbicides. Exposure to these chemicals can cause acute poisoning and chronic illnesses, including cancer, neurological disorders, and hormonal imbalances. Pesticide runoff contaminates water sources, affecting communities downstream. (2) **Loss of Nutritional Diversity:** Monoculture promotes a narrow range of staple crops (e.g., wheat, rice, corn), reducing the availability of diverse and nutrient-rich foods. This can lead to malnutrition, micronutrient deficiencies (e.g., Vitamin A, iron), and associated diseases such as anemia and scurvy. (3) **Emergence of Zoonotic Diseases:** Large-scale monoculture reduces biodiversity, disrupting natural predators and ecological balances. The proximity of monoculture farms to animal farming operations can facilitate the spread of zoonotic pathogens (e.g., avian influenza, COVID-19). (4) **Increased Antibiotic Resistance:** Monoculture's pest problems often extend to livestock, leading to the overuse of antibiotics in nearby animal farms to prevent disease. This contributes to the global crisis of antibiotic resistance, making human infections harder to treat. (5) **Air and Water Pollution:** Monoculture practices often involve burning crop residues, releasing pollutants like particulate matter and greenhouse gases into the air, causing respiratory diseases. Fertilizer runoff contaminates drinking water with nitrates, leading to conditions like blue baby syndrome. (6) **Mental Health Impacts on Farmers:** The economic instability associated with monoculture (e.g., crop failures due to pests or market price fluctuations) contributes to stress, depression, and higher suicide rates among farmers.

Poverty and Indebtedness: Monoculture farming, while profitable in the short term, often leads to poverty and indebtedness among farmers, especially smallholders. This is due to its vulnerability to market fluctuations, environmental risks, and increased dependence on external inputs. Over time, monoculture undermines economic stability, perpetuating cycles of poverty. How Monoculture Causes Poverty and Indebtedness: [1] **Market Dependency and Price Fluctuations**: Farmers growing a single crop are highly dependent on market prices for that crop. Global overproduction or market shocks can lead to sharp price drops, leaving farmers unable to recover costs. (2) **High Input Costs**: Monoculture requires heavy use of chemical fertilizers, pesticides, and herbicides. These costs, combined with expensive seeds (e.g., genetically modified or hybrid seeds), force farmers to borrow money, often at high interest rates. (3) **Crop Failure Risks**: Pests, diseases, or adverse weather conditions can devastate monoculture crops, leading to total income loss. Without crop diversity, farmers have no fallback income source. (4) **Depletion of Natural Resources**: Continuous monoculture depletes soil fertility and groundwater, increasing the need for costly inputs like synthetic fertilizers and irrigation systems. These additional expenses exacerbate financial strain. (5) **Dependence on Credit**: Farmers often take loans to purchase inputs or recover from crop failures. Repeated borrowing without adequate returns leads to debt cycles. (6) **Lack of Resilience**: Monoculture systems are less resilient to climate change, making them prone to frequent losses due to droughts, floods, or temperature changes. (7) **Middlemen Exploitation**: Small-scale farmers in monoculture systems often depend on middlemen to sell their crops, who pay low prices and take a large share of the profits.

1.3.2 Polyculture:

Polyculture is a mixed cropping system which includes more than two crops at a time and place and integrates farm animals. Polyculture systems of the world can be divided into two types (1) traditional agricultural landscape and (2) modern ecological agriculture. Here are a few examples:

Traditional Agricultural Landscape (Agricultural Heritage):

Baranaja (twelve grains, India), Bark Garden (Papua New Guinea), Boulder Farms (Ireland), Bunong Shifting Cultivation (Cambodia), Caatinga Agroforestry (Brazil), Chinampas (Mexico), Dehesa (Spain, Portugal), Dyak Farming (Borneo), Fazenda (Brazil), Hortillonnages (France), Huerta (Spain), Jhum (India), Kampong (Indonesia, Malaysia, Philippines), Masia (Spain), Milpa (Three Sisters, Mexico), Pekarangan (Indonesia), Polyculture (Madagascar), Pusaka Agroforestry (Indonesia), Rooibos Cultivation (South Africa), Satoyama (Japan), Sorjan Farming (Indonesia), Subak System (Bali, Indonesia), Talaiot Terraces (Spain), Terra Preta (Amazonian Basin), Wadi Agriculture (Yemen), Waru-Waru (Peru, Bolivia), Zabo Cultivation (India), Zai (Burkina Faso, Mali, Niger), Ziro Valley (India) and

Vavilov Centres:

Vavilov Centers are regions identified by Russian geneticist Nikolai Vavilov as centers of origin for cultivated plants. These areas are rich in genetic diversity and are where many domesticated crops first originated. Vavilov's work laid the foundation for understanding the geographic and genetic origins of agriculture. It needs to establish more such Vavilov Centres to cover all food crops of the world. Key Features of Vavilov Centers: (1) **High Genetic Diversity**: These centers are hotspots for a wide variety of plant traits, including those valuable for breeding, such as disease resistance, yield potential, and environmental adaptability. (2) **Centers of Domestication**: These regions are where humans first began cultivating and domesticating wild plants. (3) **Adaptation** to Local Environments: Crops in these centers are well-adapted to diverse climates, soils, and ecological conditions. (4) **Cultural and Historical Significance**: The development of agriculture in these areas influenced the rise of ancient civilizations. The 8 Primary Vavilov Centers are: (1) **Central America and Mexico**: Crops: Maize, beans, squash, chili peppers, cotton, avocado, sweet potato. (2) **Andean Region (South America)**: Crops: Potatoes, quinoa, cassava, peanuts, coca. (3) **Mediterranean**: Crops: Wheat, barley, olives, grapes, lentils, chickpeas. (4) **Middle East (Fertile Crescent)**: Crops: Wheat, barley, peas, lentils, figs. One of the earliest centers of agriculture, supporting ancient Mesopotamian cultures. (5) **Ethiopia**: Crops: Coffee, teff, sorghum, millet. (6) **Central Asia**: Crops: Wheat, barley, rye, grapes, apples, pistachios. Key region for temperate-zone crops. (7) **South Asia (India and Surrounding Regions**): Crops: Rice, sugarcane, eggplant, mango, black pepper, turmeric. Major contributor to global food staples. (8) **East Asia (China)**: Crops: Rice, millet, soybean, tea, bamboo.

Globally Important Agricultural Heritage Systems (GIAHS):

Globally Important Agricultural Heritage Systems (GIAHS) are unique agricultural systems recognized by the Food and Agriculture Organization (FAO) for their remarkable biodiversity, traditional knowledge, and cultural heritage. These systems have evolved over centuries, demonstrating sustainable practices that balance human needs and environmental conservation. **Objectives of GIAHS**: (1) **Conservation**: Protect agricultural biodiversity, traditional knowledge, and ecosystems. (2) **Sustainability**: Promote sustainable farming practices that ensure food security. (3) **Cultural Preservation**: Safeguard traditional farming communities and their way of life. (4) **Development**: Support eco-tourism, local economies, and sustainable livelihoods. **Key Characteristics of GIAHS:**

(1) **Agro-Biodiversity:** High genetic diversity of crops, livestock, and associated species. (2) **Traditional Knowledge Systems:** Indigenous knowledge of farming techniques, water management, pest control, and seed saving. (3) **Resilient Ecosystems:** Farming systems that adapt to changing environmental and socio-economic conditions. (4) **Cultural Heritage:** Rich cultural traditions, rituals, and festivals linked to agriculture. (5) **Food and Livelihood Security:** Provide sustainable livelihoods and contribute to local and global food security. **Examples of GIAHS:** (1) **Rice-Fish Farming System (China):** Integrated system where fish are raised in flooded rice paddies, enhancing pest control, nutrient recycling, and productivity. Location: Zhejiang Province, China. (2) **Ifugao Rice Terraces (Philippines):** Ingenious terrace farming system carved into mountains, supported by intricate water management techniques. Location: Luzon, Philippines. [3] **Traditional Qanat Irrigation System (Iran):** Underground irrigation channels that distribute water efficiently in arid regions. Location: Yazd Province, Iran. (4) **Andean Agriculture (Peru and Bolivia):** Diverse cropping systems on terraces, growing potatoes, quinoa, and maize with traditional techniques. Location: Andes Mountains, South America. (5) **Chilika Lagoon Fisheries (India):** Sustainable fishing practices in the largest brackish water lagoon in Asia, supporting biodiversity and livelihoods. Location: Odisha, India. (6) **Alpujarra Agricultural System (Spain):** Terraced fields and water channels developed during Moorish rule, combining Mediterranean crops like olives, grapes, and almonds. Location: Andalusia, Spain.

Modern Ecological Agriculture (Invented by Individual entrepreneurs):

Agroecology (Miguel A. Altieri), Agroecological Symbiosis (Jyrki Luukkanen), Agroforestry (John Bene), Analog Forestry (Dr. Ranil Senanayyake), Biointensive Agriculture (Alan Chadwick),

Conservation Agriculture (Rolf Derpsch), Ecological Farming (Sir Albert Howard), Fertility Farming (Edward H. Faulkner), Food Forest (Robert Hart), Forest Farming (J. Russell Smith), Forest Gardening (Robert Hart), Korean Natural Farming (KNF, Cho Han-Kyu), Natueco Farming (Dr. Shipad A. Dabholkar), Natural Farming (Masanobu Fukuoka), No-Dig Gardening (F. C. King), No-Till Farming (Edward H. Faulkner), Permaculture (Bill Mollison, David Holmgren), Regenerative Agriculture (J. I. Rodale), Rewilding Agriculture (Chris Baines), Sustainable Intensification (Jules Pretty), Syntropic Agriculture (Ernst Gotsch), Taungya Cultivation (Henry Ward), Vedic Agriculture (Pratap C. Aggarwal), Zero Budget Natural Farming (ZBNF, Subhas Palekar).

Characteristics of Polyculture:

The Holocene: The Holocene is the current geological epoch, beginning around 11,700 years ago, following the Last Glacial Period. It is characterized by significant human development, including the rise of agriculture and civilizations. During the period of the Holocene, polyculture was the only form of agriculture.

Ecological Agriculture (Natural Intelligence): Indigenous people applied their natural intelligence to decode the functions (ecosystem services) of the natural forest ecosystems to design the most prolific design of artificial (man-made) ecosystems which closely mimic old growth forests to provide maximum ecosystem services to man, which is now known as polyculture agroecosystem. Critical analysis of these traditional agroecosystems reveal that they indeed follow the principles of ecology, so they are modern ecological agriculture.

Biodiversity Based Agriculture: Biodiversity is the diversity of life on earth. Polyculture systems efficiently manage biodiversity in its agroecosystem, to maximise the ecosystem services of crop production, crop protection and profit maximisation of the farmers.

Do–Nothing Agriculture / Minimalist Agriculture: Polyculture relies on complete automation of ecosystem services, for optimising crop production and crop protection. Farmers role in polyculture is limited to selection of crop plant varieties and animal breeds, optimisation of the portfolio size and design of the farm landscape and multilayer polycropping system. The agroecosystem will function automatically forever, to provide multiple sources of passive income to the farmer.

Complex Agroecosystem: Polyculture agroecosystems are highly complex ecosystems, comprising multiple crops, multiple farm animals and multiple aquatic animals, along with their complex interspecific interactions and population dynamics.

Knowledge Based Agriculture: Polyculture is basically a knowledge based agriculture. Farmers should have traditional holistic knowledge of agriculture and allied activities as well as modern scientific knowledge on ecology, to become successful in polyculture. Traditional agricultural knowledge is oral, personal and generational, which is not available in books or peer-reviewed journals. Irrespective of their level of education, individual farmers have to play multiple roles in their own polyculture agribusiness such as economic botanist, plant breeder, plant pathologist, entomologist, agronomist, accountant, research analyst, marketing manager, public relation manager and general manager.

Holistic Knowledge: Holistic knowledge is an understanding that considers the interconnections and relationships between all components of a system, rather than focusing solely on individual parts in isolation. It is rooted in the idea that systems—whether they are ecosystems, societies, or human bodies—function as a whole, where the entirety is greater than the sum of its parts. **Characteristics of Holistic Knowledge: Interconnectedness:** Recognizes that all elements in a system influence and depend on one another. **Multidisciplinary:** Draws from diverse fields or disciplines to provide a comprehensive understanding. **Systems Thinking:** Emphasizes relationships, patterns, and the dynamics of the whole system. **Sustainability:** Aims to balance and maintain the health of systems over the long term. **Contextual Understanding:** Considers the broader context, including cultural, environmental, and historical factors.

Nature Based Solution (NbS): Nature Based solutions are those solutions which use living organisms such as plants, animals and microorganisms (biology, ecology), instead of using chemicals (chemistry) or machines (physics) for solving the problems of mankind. Common methods of developing nature based solution include: (1) rating and ranking of plants, animals and microorganisms of the world, on the basis of their capacity to provide various ecosystem services to man, (2) domestication and breeding of the top rated plants, animals and microorganisms, to increase their efficiency to provide ecosystem services, (3) designing the best artificial (man-made) ecosystems, such as agriculture, garden, park, sanctuary, agroforestry or forest, with the top rated plants, animals and microorganisms, to maximise ecosystem services, sustainability and resilience.

Sustainable Agriculture: Sustainable agriculture is a farming approach that focuses on producing food, fiber, and other agricultural products in a way that maintains the ecological balance, promotes economic viability, and ensures social equity for current and future generations. It emphasizes the responsible use of natural resources while protecting the environment and supporting the well-being of farming communities.

Climate Resilient Agriculture: The capacity of agroecosystems to resist or recover from environmental disturbances while maintaining their essential functions is known as resilience. Natural farming agroecosystems adapt to changing climatic conditions, mitigate greenhouse gas emissions, and sustain productivity and profitability through genetic crop improvement, crop diversification in mixed cropping systems, improvement of soil health and plant health, water cycle management and integration of trees in the agroecosystem. This approach emphasizes resilience to climate variability, extreme weather events, and long-term changes like rising temperatures and unpredictable rainfall.

Economic Democracy: Polyculture facilitates economic democracy in several ways: (1) **Decentralization of Control:** Polyculture encourages small-scale, diverse farms over large-scale monocultures. This decentralization reduces dependency on large agribusiness corporations and empowers individual farmers, fostering economic equality and local decision-making. (2) **Resilience and Sustainability:** By cultivating multiple crops, polyculture systems are less vulnerable to pests, diseases, and market fluctuations. This stability allows smallholders and communities to sustain their livelihoods, ensuring more equitable economic opportunities. (3) **Employment Generation:** Polyculture requires more intensive management, which creates jobs in rural areas. This leads to a more equitable distribution of economic benefits across communities. (4) **Local Market Development:** Diverse produce from polyculture farms supports local markets by offering a range of products. This helps small-scale farmers capture a fair share of market value, reducing economic disparity. (5) **Community-Based Economies:** Polyculture encourages cooperative farming and collective decision-making. Communities can work together, pool resources, and share profits, which strengthens local economies and supports democratic principles. (6) **Reduced Dependence on External Inputs:** Polyculture systems often use natural farming techniques, reducing reliance on costly inputs like synthetic fertilizers and pesticides. This ensures that small farmers retain more of their income, promoting economic fairness.

No Leverage: Polyculture is traditionally practiced by the indigenous people and a few conscious small-scale farmers on their own interest and motivation. Polyculture farmers do not get any financial or technical support from the government or any other organisations to increase their productivity or profitability. Farmers themselves have to bear all the costs of self-education, personal research and personal investment in polyculture.

Objectives of Polyculture:

The main objectives of modern polyculture should include:

To Maximise Profit of the Farmers: Polyculture should be considered as a normal business of the farmers and the primary objective of this business should be profit maximisation for the farmers. Polyculture systems can minimise the risk and cost of production of crops and livestocks and can maximise the aggregate yield, quality of products and price of the products, to maximise profit of the farmers.

To Leverage Technology to Maximise Profit of the Farmers: In modern context, farmers and entrepreneurs should leverage modern technology such as social media platforms, artificial intelligence, machine learning, data science and blockchain technology (1) to optimise and rebalance the plant and animal portfolio of the natural farming agroecosystem, (2) to biomonitor soil health, plant health, natural control of insect pests and diseases and to biomonitor the environment health, (3) to monitor carbon sequestration in soil, plants and animals of the natural farming agroecosystem to calculate carbon credit entitlement, and (4) to multiply the sources of income through direct marketing, digital marketing through social networking sites.

To Maximise Rainwater Harvesting and Conservation: Water is the most important natural capital required for polyculture.

Traditional polyculture systems of the world have developed many unique water harvesting structures according to their own unique situations to save water for agriculture and animal husbandry.

To Improve Soil Health: Restoration and conservation of soil health is the primary concern of polyculture, to maximise crop production and crop protection. Polyculture improves soil health solely through efficient water management, agrobiodiversity management and agroecosystem management. For example, earthworms maintain soil tilth (a process known as bioturbation), soil microorganisms decompose organic matter and maintain soil aggregates, plant, animal and microorganism biodiversity maintains water cycle and nutrient cycle in soil. In this way, soil organisms continuously improve soil health.

To Improve Plant Health and Immunity: Polyculture systems intend to maintain plant health and plants immunity against insect pests and diseases, automatically, through improvement of soil health and improvement of interspecific interactions among plant, animal and microorganism biodiversity in the agroecosystem (natural control). For example, earthworms and bacteria decompose and mineralise organic biomass into mineral nutrients for uptake by plants, arbuscular mycorrhizal fungi (AMF) and plant growth promoting rhizobacteria (PGPR) form symbiotic association with plant roots, to provide water and nutrients to the plants, at optimum composition, dose and timing, that ensures healthy growth, development and reproduction in plants. AMF and PGPR also induce plant defence and plant immunity against insect pests and diseases, which further improve plant health.

To Improve Animal Health and Immunity: Polyculture systems intend to improve animal health and immunity against diseases and pests, automatically, through natural feed, balanced nutrition, natural control of pests, feeding medicinal plants and free-range movement of livestock, ultimately to minimise the cost of rearing of farm animals and to minimise the cost of medicine.

To Ensure Nutritional Security of the Consumers: Small-scale farmers practice polyculture to raise multiple types of crops in their field, throughout the year. This ensures balanced nutrition for their family members without incurring any cost. Farmers also provide opportunity to the consumers to buy diverse, chemical-free, healthy and nutritious foods, directly from the farmers, throughout the year for their balanced nutrition.

To Improve Environment Health: Polyculture systems automatically improve soil health, water health, plant health, animal health, human health and ultimately environment health (also known as one health).

To Mitigate climate change: Polyculture agroecosystems are nature based solutions (NbS) for adaptation and mitigation of climate change and extreme weather events. Polyculture improves carbon sequestration in soil, plants, and animals, and significantly reduces use of fossil fuel and emission of greenhouse gases in the atmosphere such as carbon dioxide, methane and nitrous oxide, from agricultural fields.

To Restore biodiversity: Polyculture systems completely rely on biodiversity for maximising crop production and crop protection therefore conservation of biodiversity is their first priority. Polyculture does not discriminate between harmful or beneficial organisms. Polyculture does not recognise the concepts of pests (harmful insects), pathogens (disease causing microorganisms) or weeds (unwanted plants). All these organisms are considered as the integral part of the agroecosystem and their complex interactions are important and essential for successful natural farming. Biodiversity is nothing but a big food web or food pyramid. For example, insect pests themselves are the food for the predators and parasitoids and without pests there will be no predators.

Reforestation: Polyculture systems often integrate trees (tree crops and forest trees) in the agricultural landscape, which effectively help reforestation of landscape.

Longevity Conservation: Inclusion of perennial plants, grafted plants, forest trees and very old plants in the natural farming agroecosystems help conservation of longevity genes and help adaptation and mitigation to climate change and biodiversity loss.

To Reduce Environmental pollution: Polyculture systems automatically reduce environmental pollution. Polyculture helps recycling of agricultural wastes within its own agroecosystem that reduces pollution in atmosphere, soil and water. Polyculture promotes local agrifood systems which significantly reduce packaging, transport, storing, cold storing, food processing and accumulation of animal waste and food waste in garbage dumps, which reduces pollution.

To Promote Sustainability and Resilience in Agriculture: Polyculture promotes sustainable and resilient agriculture because it takes care of the soil health, environment health and farmers income growth.

To Promote Biovillage: A biovillage is a community-focused initiative aimed at promoting sustainable living through ecological practices. It emphasizes sustainable agriculture, biodiversity conservation, and the use of renewable resources. Bio-villages serve as models for sustainable rural development, enhancing food security and community resilience while addressing climate change challenges (Kesavan PC, Swaminathan MS. 2020).

To Promote Eco-Agritourism: Eco-agritourism combines agriculture and tourism, focusing on sustainable practices that benefit both the environment and local communities. It promotes rural culture and offers visitors experiences like farm stays, organic farming, and local food festivals. This form of tourism enhances economic viability for small farms while fostering community engagement and environmental stewardship. By attracting urban tourists seeking authentic rural experiences, eco-agritourism supports local economies, preserves agricultural heritage, and encourages sustainable agricultural practices, making it a vital component of modern rural development strategies.

To promote Bioeconomy: The bioeconomy refers to the economic activities centered on the production, utilization, and conservation of biological resources, aiming for sustainability across various sectors. It encompasses agriculture, forestry, fisheries, and the development of bio-based products and processes. The bioeconomy promotes renewable resources to reduce reliance on fossil fuels, enhance food security, and foster innovation through biotechnology. It aligns with principles of the circular economy by emphasizing reuse and recycling to minimize waste. Ultimately, the bioeconomy seeks to balance economic growth with ecological sustainability, addressing global challenges like climate change and resource depletion.

To Promote Circular Economy: The circular economy is an economic model that contrasts with the traditional linear approach of "take, make, dispose." It focuses on sustainability through three core principles: **Design out waste and pollution:** Products are designed to minimize waste and environmental impact from the outset. **Keep products and materials in use:** Emphasizes reuse, repair, and recycling to extend the lifecycle of resources. **Regenerate natural systems:** Encourages practices that restore and enhance natural ecosystems. This model aims to create a closed-loop system that maximizes resource efficiency, reduces environmental harm, and fosters economic resilience.

To Achieve Sustainable Development Goals (SDGs): The Sustainable Development Goals (SDGs) are a set of 17 global objectives established by the United Nations in 2015, aiming to address interconnected challenges such as poverty, inequality, climate change, environmental degradation, and peace. Each goal includes specific targets—169 in total—measured by 232 indicators to track progress. Key goals include: **End poverty** in all forms (SDG 1), **Achieve gender equality** (SDG 5), **Combat climate change** (SDG 13). The SDGs apply universally to all countries and emphasize the principle of "leaving no one behind," promoting an inclusive approach to sustainable development.

Methods of Polyculture:

Plant Domestication and Plant Breeding: Indigenous people and traditional farmers practice decentralised systems of plant domestication and plant breeding for thousands of years, to sustain their own diversified systems of polyculture.

Animal Breeding: Indigenous people have domesticated all the farm animals more than 10,000 years ago and they bred a number of animal breeds for their use as meat, milk or egg production.

Seed Preservation and Seed Treatment: Traditional knowledge of seed quality control, seed preservation and seed treatment were evolved among indigenous communities thousands of years ago. The methods include drying, seed coating, and preservation in earthen pots.

Water Harvesting and Water conservation: The methods of rainwater harvesting and water conservation were evolved in ancient civilizations,thousands of years ago.

No-Till: No till is the most effective method of improvement of soil health and plant health. Because no-till restores soil horizons, soil structure, water holding capacity of soil, soil organic carbon and soil biodiversity.

No Off-Farm Input: Avoiding all sorts of commercial chemical inputs (hybrid or GMO seed, fertilisers, micronutrients, plant growth regulators, herbicides, insecticides, fungicides and antibiotics) in agricultural fields have multiple benefits in conservation of biodiversity, mitigation of climate change and profit maximisation of the farmers.

Probiotics: Farmers often develop crude microbial formulation through fermentation of locally available organic materials such as animal dung and urine, kitchen waste, dairy waste, farm waste and native soil inoculant, for their application in soul or on plant for improvement of soil health and plant health. These are known as probiotics.

Mulching: Mulching is a traditional practice of maintaining a thick layer of organic materials, such as dry leaves, on the soil surface throughout the year, to improve crop production and crop protection. Mulching improves soil water, soil organic matter, soil biodiversity, soil health, plant health and weed control.

Cover Cropping: Cover cropping is a traditional practice of maintaining green plants in the field throughout the year to keep the soil always covered with green leaves. Cover Cropping improves soil water, soil organic matter, soil biodiversity, soil health snd weed control. In addition, legume cover crops, in symbiotic association with rhizobium bacteria, add nitrogen to the soil.

Multilayer Polycropping: Multilayer polycropping is a traditional cropping system, where multiple crops are grown in a mixed cropping style, and crops are stacked in multiple layers of canopies, according to their heights to maximise the cropping Intensity and to minimise the risk of crop loss due to climate change.

Beekeeping: Beekeeping is a common practice among indigenous farmers to maximise biotic pollination of the pollination sensitive crops by native pollinator biodiversity as well as to produce honey, a profitable agricultural product.

Mixed Animal Husbandry: Indigenous farmers practice integrated farming of crops and livestock, to effectively recycle agricultural wastes as animal feed and to recycle animal waste as organic manure for crops. Mixed rearing of cattle and poultry birds help natural control of both agricultural pests and animal pests by poultry birds as well as poultry birds turn the compost pit to help good decomposition of organic matter.

Mixed Aquaculture: Each polyculture farm should contain a freshwater pond to supply drinking and bathing water to farm animals, birds and honeybees. Farmers should practice mixed aquaculture with fishes, shrimps, crabs, turtles, frogs, oysters, snails and slugs, to maximise their profit.

Direct Marketing: The existing centralised marketing system for monoculture is completely useless for marketing of polyculture products. There is no mechanism for fair price discovery of polyculture products. Farmers should adopt the practices of direct marketing, digital marketing and eco-agritourism to assert their rights to fix the price of their own products.

CHAPTER 2

MODEL BUSINESS PLAN FOR NATURAL FARMING AGRIBUSINESS

Natural farming can be a highly profitable business and preparing a formal business plan is the primary action for maximising profit for the small-scale farmers.

2.1 Executive Summary:

Ideal Natural Farm is a private limited company dedicated to revolutionizing Indian agriculture through sustainable and natural farming practices. By focusing on ecological balance, soil health, and profitability, the company aims to deliver high-quality, chemical-free produce to consumers while enhancing farmers' livelihoods.

Vision: To be a leader in promoting natural farming as a profitable and sustainable alternative in agriculture.

Mission: To produce premium-quality, chemical-free agricultural products, educate farmers, and create an ecosystem where nature and agriculture thrive together.

2.2 Business Objectives:

(1) Establish a scalable model of natural farming for high-value crops. (2) Develop a direct-to-consumer supply chain for fresh and processed agricultural produce. (3) Provide training programs to farmers on natural farming practices. (4) Build partnerships with retail chains, restaurants, and e-commerce platforms. (5) Achieve profitability within three years of operations.

2.3 Market Analysis:

Industry Overview: (1) Growing consumer demand for chemical-free, organic products. (2) Government incentives and subsidies for natural farming. (3) Increased awareness of the environmental benefits of sustainable practices.

Target Market (1) Urban middle- and high-income families seeking organic produce. (2) Health-conscious consumers and specialty retailers. (3) Export markets for premium-quality organic goods.

Competitive Analysis:

Competitors: Large organic brands and small-scale natural farming cooperatives.

Differentiation: Lower input costs, superior product quality, farmer education programs, and ecological sustainability.

2.4 Business Strategy:

Product Offering: (1) Primary Produce: Chemical-free fruits, vegetables, grains, and pulses. (2) Processed Goods: Natural oils, pickles, flours, and spices. (3) Services: Farmer training, consultancy, and soil health improvement programs.

Revenue Streams: (1) Sale of produce and processed products. (2) Revenue from training programs and workshops. (3) Export contracts for high-value crops.

2.5 Operational Plan:

(1) Land Acquisition and Farming. (2) Identify fertile agricultural land in regions with high potential for natural farming. (3) Cultivate high-value crops such as spices, grains, and horticultural produce.

Production Techniques: (1) Use techniques like mulching, crop rotation, and natural pest management. (2) Employ natural fertilizers (compost, cow dung) and bio-pesticides.

Logistics and Distribution: (1) Develop cold chain storage and transport facilities for perishable items. (2) Partner with e-commerce platforms and retail stores for urban delivery.

2.6 Marketing Plan:

Brand Positioning: (1) Position NatureLab Agro as a trusted provider of premium, chemical-free agricultural products.

Marketing Channels: (1) **Digital Presence:** Website, social media, and influencer partnerships. (2) **Offline Campaigns:** Farmer markets, exhibitions, and organic food festivals. (3) **Collaborations:** Tie-ups with restaurants, wellness centers, and supermarkets. (4) **Certifications:** Obtain certifications for organic and natural farming to enhance credibility.

2.7 Financial Plan:

Initial Investment: (1) Land acquisition and preparation. (2) Equipment and tools for natural farming. [3] Marketing and branding expenses.

Projected Revenue (3 Years):

(1) Year 1: INR 20 Lakhs

(2) Year 2: INR 50 Lakhs

(3) Year 3: INR 1 Crore

Funding Sources:

(1) Founder's capital.

(2) Bank loans or government grants for sustainable agriculture.

(3) Private equity or angel investment in the agribusiness sector.

2.8 Risk Analysis:

(1) **Market Risks**: Uncertainty in consumer demand for organic produce. Mitigation: Diversify product range and focus on processed goods.

(2] **Climate Risks**: Impact of unpredictable weather on crop yield. Mitigation: Invest in resilient farming techniques and insurance.

(3) **Pest Risk**: Impact of incidence of insect pests and diseases. Mitigation: Design appropriate plant biodiversity for natural control. (4) **Operational Risks**: Resistance from traditional farmers. Mitigation: Offer practical demonstrations and financial incentives.

2.9 Sustainability and CSR:

(1) Commit to zero-waste farming practices. (2) Work with local communities to ensure economic upliftment. (3) Advocate for policies promoting natural farming.

CHAPTER 3

DE NOVO DOMESTICATION OF PLANTS, ANIMALS AND MICROORGANISMS FOR NATURAL FARMING

De novo domestication involves the intentional transformation of wild species into domesticated varieties to meet human needs, particularly in agriculture. This process can utilize traditional breeding methods, aiming to introduce desirable traits rapidly compared to historical domestication, which took millennia. It encompasses four main steps: selecting starting materials, establishing transformation systems, editing genes, and evaluating new cultivars. De novo domestication is seen as a solution for enhancing crop resilience and sustainability amid climate challenges (Zhang J et al. 2023).

Domestication Syndrome:

Domestication syndrome in plants refers to the set of traits that distinguish domesticated plants from their wild ancestors. These traits arose through human selection for characteristics that improve the plant's utility, productivity, and ease of cultivation. Domestication has fundamentally altered plant morphology, reproduction, and physiology to suit human needs. Common traits of domestication syndrome in plants: (1) **Seed and Fruit Characteristics: Loss of seed dispersal mechanisms:** Wild plants often have mechanisms to scatter seeds, such as shattering grains in cereals. Domesticated plants, like wheat and rice, retain seeds on the plant for easier harvesting. **Increased seed or fruit size:** Domesticated plants tend to have larger seeds or fruits compared to their wild ancestors (e.g., corn vs. teosinte, modern apples vs. wild apples). **Reduced seed dormancy:** Domesticated seeds germinate more uniformly and predictably, unlike wild seeds, which may remain dormant to survive adverse conditions. (2) **Growth and Productivity: Compact growth forms**: Domesticated plants often have reduced branching or a more uniform growth pattern (e.g., dwarf varieties of wheat and rice). **Higher yield**: Selection has favored plants that produce more edible parts (e.g., larger ears of corn, more grains in wheat). (3) **Ease of Harvest: Simultaneous ripening**: Domesticated crops often ripen uniformly to facilitate harvesting. **Non-toxic traits**: Many wild plants contain toxins or compounds that deter herbivory. Domesticated varieties have reduced levels of these (e.g., sweet almonds vs. bitter almonds). (4) **Reproductive Changes: Shift from outcrossing to self-pollination**: Domesticated plants like wheat and barley are often self-pollinating, which ensures uniform traits in crops. **Loss of natural defenses**: Reduced thorns, spines, or tough seed coats (e.g., seedless fruits like bananas). (5) **Aesthetic and Culinary Traits: Improved taste and texture**: Domestication has often prioritized sweeter, less bitter flavors (e.g., modern carrots vs. their bitter ancestors). **Diversified appearances**: Human selection has created diverse shapes, sizes, and colors in crops (e.g., tomatoes, peppers). **Examples of Domesticated Plants**: (1) Wheat and Barley: Loss of seed shattering and uniform ripening. (2) Corn (Maize): Enlarged cob size and more rows of kernels compared to wild teosinte. (3) Rice: Reduced seed shattering and increased grain yield. (4) Fruits (e.g., apples, bananas): Larger size, sweeter taste, and reduced seeds. (5) Vegetables (e.g., cabbage, broccoli): Selection for specific traits like larger leaves or flower heads. **Genetic Basis**: The traits of domestication syndrome are often controlled by a few key genes with large effects. For example: TB1 gene in corn controls branching and growth form. Q gene in wheat affects seed shattering and head shape.

Stages of Domestication:

The process of plant domestication can be understood in stages, transitioning from wild plants to fully domesticated crops. These stages include non-domesticated, semi-domesticated, and domesticated phases, each reflecting increasing levels of human influence on plant characteristics and cultivation. (1) **Non-Domesticated Stage**: Wild plants in their natural ecosystems, with no significant human influence. These are known as wild relatives or wild progenitor. **Traits**: Fully adapted to survive and reproduce in the wild. Natural seed dispersal mechanisms (e.g., shattering grains, wind dispersal). Seed dormancy to survive unfavorable conditions. High genetic diversity for resilience in natural environments. Often contain toxic compounds as natural defenses. **Human Interaction**: Humans gather wild plants for food, medicine, or materials but do not alter their genetics or environment significantly. Examples: Wild teosinte (ancestor of maize), wild wheat, and wild rice. (2) **Semi-Domesticated Stage**: Wild plants begin to show traits influenced by human selection, often unintentionally, as they are cultivated or managed in human-controlled environments. **Traits**: Reduced seed dispersal: Early humans may have favored plants with seeds that stayed on the plant for easier harvesting. Partial loss of seed dormancy for quicker and more uniform germination. Larger or tastier fruits and seeds due to unintentional selection during harvesting and planting. Genetic diversity remains relatively high, as wild populations still contribute to the gene pool. **Human Interaction**: Early forms of cultivation, such as clearing land, planting seeds, or selectively harvesting certain plants. Examples: Early forms of wheat, barley, and lentils during the Neolithic era. (3) **Domesticated Stage**: Plants are fully adapted to human-controlled environments, often to the point where they cannot survive or reproduce effectively in the wild without human intervention. **Traits**: Complete loss of seed dispersal mechanisms (e.g., non-shattering grains). Uniform ripening and growth for easier harvesting. Larger, more palatable fruits, seeds, or roots. Reduced natural defenses, such as thorns, bitter compounds, or tough seed coats. Reduced genetic diversity due to selective breeding. **Human Interaction**: Intensive cultivation, breeding, and propagation by humans. Dependence on human intervention for planting, weeding, and pest control. Examples: Modern maize, wheat, rice, and domesticated fruits like bananas, apples, and tomatoes.

3.1 Plant Domestication:

Theories on Plant Domestication:

Plant domestication is the process by which humans select and cultivate plants with desirable traits for agriculture, transforming wild species into crops suited for human use. Several theories explain the origins and development of plant domestication, often linked to environmental, social, and cultural factors: (1) **Natural Habitat Hypothesis** (Hilly Flanks Hypothesis, Robert Braidwood): This theory suggests that domestication occurred in areas where wild ancestors of domesticated plants were naturally abundant. (2) **Oasis Hypothesis** (V. Gordon Childe): This theory states that domestication started in oases or fertile areas during the dry climatic conditions of the late Pleistocene. (3) **Dump Heap Hypothesis** (Edgar Anderson): This theory suggests that waste disposal areas (dump heaps) became accidental nurseries for plants. (4) **Evolutionary and Intentionality Theory** (David Rindos): This theory views domestication as a co-evolutionary process where both humans and plants benefited. (5) **Cultural Progress Hypothesis**: This theory assumes that humans domesticated plants as they became more "civilized" and intelligent. (6) **Demographic Pressure Hypothesis** (Population Pressure Theory, Lewis Binford): This theory suggests that increasing human populations led to the need for more reliable food sources. (7) **Social Hypothesis** (Brian Hayden): This theory suggests that domestication was driven by social factors, such as the desire for surplus food to support feasting, trade, or status building. (8) **Climate Change Hypothesis:** This theory links plant domestication to climate stabilization after the last Ice Age (~12,000 years ago). (9) **Niche Construction Theory:** This theory views domestication as a result of humans altering ecosystems to create niches for preferred plants. (10) **Multiple-Origin Hypothesis:** This theory argues that domestication occurred independently in various parts of the world.

3.1.1 Objectives of Plant Domestication:

The objectives of plant domestication revolve around improving the utility, adaptability, and productivity of plants for human needs. These objectives include: (1) **Food Production:** Enhance the yield and quality of edible plant parts, such as fruits, grains, seeds, and tubers, to ensure a reliable food supply. (2) **Adaptability:** Modify plants to thrive in diverse climatic and environmental conditions, enabling cultivation in varied ecosystems. (3) **Ease of Cultivation:** Select traits that simplify agricultural practices, such as uniform growth, resistance to pests and diseases, and tolerance to drought or salinity. (4) **Nutritional Improvement:** Increase the nutritional value of crops by enhancing protein, vitamin, and mineral content. (5) **Storage and Shelf Life:** Develop varieties with longer storage capabilities, reduced spoilage, and improved transportability to minimize post-harvest losses. (6) **Aesthetic and Cultural Needs:** Domesticate ornamental plants for aesthetic appeal and ceremonial or cultural uses. (7) **Non-Food Uses:** Cultivate plants for industrial purposes, such as fiber, oil, medicinal compounds, and biofuel production. (8) **Uniformity and Predictability:** Ensure consistency in plant growth and development, making farming practices more predictable and manageable. (9) **Resource Efficiency:** Develop crops that use water, nutrients, and sunlight more efficiently, minimizing environmental impacts. (10) **Genetic Diversity Preservation**: Maintain genetic variability to ensure resilience against changing environmental conditions and future agricultural challenges.

3.1.2 Ancient Methods of Plant Domestication:

Ancient methods of plant domestication involved simple, observational practices and trial-and-error approaches that gradually transformed wild plants into reliable food sources. These methods, developed over thousands of years, were guided by practical needs, environmental conditions, and the inherent knowledge of early agricultural societies. Here are the key methods used in ancient plant domestication: (1) **Selection of Desirable Traits**: Early humans observed natural variations in wild plants and selected seeds from plants with favorable traits, such as: Larger seeds or fruits. Non-shattering seed heads (to prevent seeds from dispersing naturally). Better taste or reduced bitterness. Resistance to pests or drought. (2) **Seed Sowing and Cultivation**: Instead of relying solely on gathering, humans began intentionally planting seeds in fertile areas. Seeds were sown near settlements for convenience. Over time, plants adapted to human-controlled environments. (3) **Propagation by Cuttings and Vegetative Reproduction**: Plants that could reproduce asexually (e.g., tubers, bulbs, or cuttings) were propagated by planting parts of the plant, such as: Tubers (e.g., yams and potatoes). Shoots or runners (e.g., bananas). Stem or branch cuttings (e.g., figs). Example: Cassava and bananas were domesticated through vegetative propagation. (4) **Domestication by Fire Clearing**: Fire was used to clear forests and encourage the growth of certain plants in open areas. Fire promoted the growth of sun-loving plants like grains. Charred soil often became more fertile, aiding early cultivation. Example: Slash-and-burn agriculture facilitated the domestication of crops like maize in the Americas. (5) **Storage and Observation**: Early societies stored seeds for later use, leading to unintentional selection of seeds with: Long storage life. Dormancy traits suitable for delayed germination. (6) **Transplanting:** Wild plants were uprooted and transplanted closer to settlements for ease of access and observation. Example: Wild rice in China may have been domesticated by transplanting it from wetlands to controlled fields. (7) **Encouraging Growth in Managed Ecosystems:** Early humans managed ecosystems to favor the growth of certain plants by: Clearing competing vegetation. Protecting plants from herbivores. Enhancing soil fertility with natural fertilizers like ash or manure. Example: Early Native Americans created "forest gardens" with domesticated squash and beans. (8) **Trial and Error with Edibility:** Early agriculturalists experimented with wild plants to determine which parts were safe to eat or use. Processing methods, like soaking or cooking, were developed to make some plants edible. Example: Early processing of bitter wild almonds led to the domesticated sweet almond. (9) **Irrigation and Water Management:** Simple irrigation techniques, like diverting river water, allowed humans to cultivate plants in otherwise dry areas. (10) **Domestication through Pollination Control:** Although less intentional, early humans may have influenced pollination by planting crops in clusters, leading to: Natural hybridization of species. Development of traits like larger fruits or faster growth. Example: Teosinte was gradually transformed into maize in the Americas. (11) **Crop Rotation and Soil Management:** Ancient farmers rotate crops and allowed fields to lie fallow, preserving soil fertility and ensuring better yields. Example: Legumes like chickpeas and lentils were cultivated alongside grains in ancient Mesopotamia.

3.2 Animal Domestication:

Animal domestication is the process through which humans selectively breed animals for specific traits, resulting in genetic adaptations that distinguish them from their wild ancestors. This process can be categorized into three main pathways: **commensal** (e.g., dogs and cats), **prey** (e.g., sheep and cattle), and **draft** (e.g., horses and donkeys) animals. The first domesticated animal was the dog, with domestication occurring over thousands of years alongside human agricultural development (Cucchi T, Arbuckle B. 2021).

3.2.1 Objectives of Animal Domestication:

The objectives of animal domestication focus on enhancing the utility, adaptability, and productivity of animals to meet human needs. Using terminology from genetics and breeding, the key objectives include: (1) **Improved Productivity: Milk Production:** Select for high-yielding breeds to enhance milk production. **Meat Yield:** Breed animals with superior growth rates, muscle development, and feed conversion efficiency. **Egg Production:** Focus on genetic lines with higher egg-laying capacity. (2) **Adaptability to Human Needs:** Modify animals genetically and behaviorly to thrive in diverse environments, such as high altitudes, arid regions, or confined spaces. (3) **Work Efficiency:** Domesticate animals for labor, such as plowing fields, transportation, and other agricultural activities, by selecting traits like strength, endurance, and docility. (4) **Improved Reproductive Traits:** Increase fertility rates and reduce generation intervals through selective breeding, enhancing overall productivity. (5] **Disease Resistance:** Develop and select strains or breeds with natural resistance to prevalent diseases, reducing the need for external medical interventions. (6) **Behavioral Traits:** Select for tameness, reduced aggression, and sociability to make animals more manageable and compatible with human interaction. (7) **Quality Enhancement: Fiber Production:** Domesticate animals such as sheep, goats, and alpacas to produce high-quality wool, hair, or fur. **Leather:** Breed animals with desirable hide quality for the leather industry. **Draught Power:**

Enhance traits like strength and endurance in animals used for pulling loads or agricultural machinery. (8) **Nutritional Efficiency:** Select breeds that efficiently convert feed into body mass, milk, or other useful outputs, minimizing resource input. (9) **Conservation of Genetic Resources:** Maintain and propagate genetic diversity in domestic animals to safeguard against diseases, environmental changes, and genetic bottlenecks. (10) **Cultural and Social Needs:** Domesticate animals for religious, ceremonial, or recreational purposes, such as horses for cavalry or dogs for companionship. (11) **Waste Utilization:** Promote breeds capable of efficiently consuming agricultural waste and converting it into useful products like manure, milk, or meat. (12) **Aesthetic and Recreational Purposes:** Domesticate animals for their beauty, such as ornamental birds, or recreational activities, like racehorses or hunting dogs.

3.2.2 Methods of Animal Domestication:

Traditional methods of animal domestication, when described using scientific terminology, involve the systematic selection, breeding, and adaptation of wild animals into forms that align with human needs. These methods primarily focus on manipulating genetic variation, behavioral traits, and phenotypic characteristics over generations. [1] **Artificial Selection: Natural Variability Utilization:** Identifying and selectively breeding individuals with desirable phenotypic traits such as tameness, higher productivity, or disease resistance. **Directional Selection:** Focusing on specific traits, such as increased milk yield or reduced aggression, to create a population with improved performance over time. (2) **Inbreeding:** Controlled mating within a small population to fix desirable traits in the progeny. This method increases homozygosity, preserving selected traits but also necessitating careful management to avoid inbreeding depression. (3) **Outbreeding:** Mating between genetically diverse individuals or populations to introduce new genetic variations and enhance hybrid vigor (heterosis) for traits such as growth rate, fertility, or adaptability. (4) **Controlled Breeding Programs:** Early forms of selective breeding involved choosing specific animals for mating based on observable traits such as size, temperament, or productivity. This led to the gradual development of landraces and regional breeds. (5) **Domestication for Behavior:** Selecting animals that exhibit reduced flight response, increased tolerance to human presence, and social behaviors conducive to living in groups, a process referred to as behavioral adaptation. (6) **Phenotypic Plasticity:** Exploiting the ability of animals to adapt to environmental changes, allowing farmers to select individuals best suited to local climates, diets, and management systems. (7) **Natural Hybridization:** Allowing crossbreeding between wild and early domesticated populations to enhance genetic diversity and incorporate beneficial traits like hardiness or disease resistance. (8) **Culling:** Eliminating animals that exhibit undesirable traits, such as aggression or poor productivity, thereby refining the population's genetic pool over successive generations. (9) **Isolation Breeding:** Geographical or controlled physical separation of a population to prevent gene flow with wild relatives, maintaining the purity of domesticated lines. (10) **Mutation Selection:** Identifying and propagating beneficial spontaneous mutations, such as polled (hornless) traits in cattle, that improve manageability or productivity. (11) **Selective Pairing:** Matching specific males and females to achieve offspring with desired combinations of traits, optimizing genetic potential for productivity or behavior. (12) **Artificial Environment Creation:** Modifying habitats and feeding patterns to favor traits like faster growth, earlier maturation, or increased reproductive success under human-controlled conditions. (13) **Imprinting:** Leveraging the critical period in early animal development to accustom young animals to human presence and management practices, enhancing tameness and adaptability. (14) **Domestication Syndrome:** Recognizing and propagating genetic traits associated with domestication, such as reduced brain size, neotenic (juvenile-like) traits, and altered reproductive cycles, which arise naturally during domestication processes. (15) **Selection for Productivity and Utility:** Animals were traditionally domesticated for specific purposes such as milk, meat, fiber, labor, or companionship. Over time, humans unintentionally selected for genetic changes that optimized these outputs.

3.3 Microorganism Domestication:

Microorganism domestication involves the selective breeding of wild microbial species, such as bacteria, yeasts, and molds, to enhance traits beneficial for human use. This process has been crucial in food production, particularly fermentation, leading to the development of products like cheese, beer, and bread. Domesticated microbes exhibit genomic specialization due to their adaptation to stable agricultural environments, which enhances their fermentative capabilities and alters their genetic makeup. Key examples include *Saccharomyces cerevisiae* for baking and brewing and *Aspergillus oryzae* for traditional Asian fermented foods (Somerville V, et al. 2024).

3.3.1 Objectives of Microorganism Domestication for Agriculture:

The objectives of microorganism domestication for agriculture involve enhancing the productivity, sustainability, and resilience of farming systems. By leveraging the genetic and metabolic potential of microorganisms, the following goals are pursued: (1) **Soil Fertility Enhancement:** Domesticate nitrogen-fixing bacteria such as Rhizobium, Azotobacter, and Frankia to improve nitrogen availability in soils. Utilize phosphate-solubilizing microorganisms like Bacillus and Pseudomonas to increase phosphorus uptake by plants. Enhance potassium-mobilizing bacteria to make potassium more accessible to crops. (2) **Plant Growth Promotion:** Develop microbial strains like Pseudomonas fluorescens and Bacillus subtilis to produce phytohormones (e.g., auxins, gibberellins) that stimulate plant growth. (3) **Biological Nitrogen Fixation (BNF):** Use symbiotic (e.g., Rhizobium) and free-living (e.g., Azospirillum) nitrogen-fixing bacteria to reduce dependence on chemical fertilizers. (4) **Pest and Disease Control:** Domesticate biocontrol agents such as Trichoderma for fungal pathogen suppression and Bacillus thuringiensis for pest management. Use antagonistic microbes to combat nematodes, bacterial wilt, and other plant diseases. (5) **Organic Matter Decomposition:** Utilize microorganisms like Actinomycetes and Aspergillus for breaking down organic residues and accelerating composting processes. (6) **Stress Tolerance Improvement:** Domesticate microorganisms that confer drought, salinity, and temperature stress tolerance to crops by producing osmolytes or other protective compounds. Examples include mycorrhizal fungi and halotolerant bacteria. (7) **Carbon Sequestration and Soil Health:** Use microorganisms like cyanobacteria and mycorrhizal fungi to enhance soil carbon sequestration, improving soil structure and fertility. (8) **Weed Suppression:** Domesticate allelopathic microorganisms that produce natural herbicides to suppress weed growth and reduce reliance on chemical herbicides. (9) **Symbiotic Interactions:** Optimize plant-microbe symbioses, such as arbuscular mycorrhizal fungi (AMF) for improved nutrient uptake and root development. (10) **Biofertilizer Production:** Scale up the production of microbial biofertilizers to provide eco-friendly alternatives to chemical fertilizers, reducing environmental impacts. (11) **Disease Resistance Induction:** Domesticate microorganisms that trigger systemic acquired resistance (SAR) or induced systemic resistance (ISR) in plants, enhancing their ability to fight pathogens. (12) **Nutrient Recycling:** Develop microbial consortia that recycle nutrients from agricultural waste, making them available for plant uptake. (13) **Reduction of Greenhouse Gas Emissions:** Use methanotrophic bacteria to reduce methane emissions in paddy fields and other agricultural systems.

(14) **Improved Crop Productivity:** Select microorganisms that increase crop yield and quality by improving nutrient use efficiency and supporting healthier plants. (15) **Environmentally Sustainable Agriculture:** Replace chemical inputs with domesticated microbial solutions to minimize soil and water pollution while promoting ecological balance.

3.3.2 Methods of Microorganism Domestication:

Traditional methods of microorganism domestication for agriculture involve practices that have been developed over centuries, often based on empirical knowledge, observation, and trial-and-error techniques. These methods focus on selecting, cultivating, and utilizing naturally occurring microorganisms to benefit agricultural productivity, soil health, and plant growth. Key traditional methods include: (1) **Natural Fermentation:** Farmers have long used fermentation to domesticate microorganisms for agricultural purposes, such as in the production of biofertilizers and organic matter breakdown. This includes the fermentation of plant residues to create compost, where microorganisms like fungi, bacteria, and actinomycetes naturally decompose organic matter into nutrient-rich humus. (2] **Composting:** Compost Teas and Cultures: Traditional farmers often encouraged the growth of beneficial microorganisms through the use of composted organic matter. Microbial populations were selected and enriched by soaking compost in water to create "compost tea," which was then applied to plants to improve growth and provide natural nutrients. (3) **Seed Inoculation:** Rhizobium Inoculation: In the past, farmers often used traditional methods to inoculate seeds with naturally occurring nitrogen-fixing bacteria, such as Rhizobium or Azotobacter, to promote better crop yields. These methods were typically performed by applying soil or compost containing beneficial bacteria to seeds before planting. (4) **Indigenous Knowledge of Symbiotic Relationships:** Mycorrhizal Fungi and Legumes: Indigenous and traditional agricultural practices have long utilized symbiotic relationships between plants and microorganisms. For example, the relationship between legumes and nitrogen-fixing rhizobial bacteria has been recognized for centuries. Similarly, traditional practices may involve encouraging the growth of mycorrhizal fungi to help plants access nutrients from the soil. (5) **Use of Animal Manure:** Microbial Inoculants in Animal Manure: Traditional farmers often relied on animal manure, which naturally contains a rich population of microorganisms, as a fertilizer. Manure was often left to ferment or was mixed with compost to further enrich microbial activity and promote the growth of beneficial soil organisms. (6) **Traditional Biocontrol Practices:** Use of Beneficial Microorganisms: Traditional farming systems often used naturally occurring microorganisms to manage pests and diseases. For instance, farmers would encourage the growth of Trichoderma spp. and Bacillus thuringiensis for the suppression of soil-borne fungal pathogens or insect pests, respectively. These methods were largely based on observation and experience rather than scientific experimentation. (7) **Agroecosystem Diversification:** Polyculture: In traditional agriculture, the diversity of plant species in polycultures and crop rotation systems helped promote beneficial microbial populations. By growing a variety of plants, farmers created a more stable environment for beneficial microorganisms to thrive, enhancing soil fertility and pest resistance. (8) **Soil Amendments:** Application of Bioactive Organic Matter: Farmers applied organic materials such as plant residues, manure, and decaying matter to soils to enrich them with microorganisms that could decompose organic matter and improve nutrient availability. These materials often contained naturally occurring beneficial bacteria and fungi. (9) **Natural Inoculation:** Inoculating with Local Microbial Strains: Traditional farming often relied on microorganisms that were indigenous to the local environment. For example, local soils and composts were used to inoculate crops with naturally occurring microorganisms, which were better adapted to the local climate and ecosystem. (10) **Use of Indigenous Microorganism (IMO) Techniques**: Korean Natural Farming (KNF): A traditional method used in East Asia, particularly in Korea, is the use of Indigenous Microorganisms (IMO) to support soil health and plant growth. Farmers collect native microorganisms from the local environment (such as from the forest floor or local fields), culture them in sugar or rice media, and then apply them to crops and soil to promote microbial diversity and improve nutrient cycling. (11) **Indigenous Fermentation of Biofertilizers:** Use of Biofertilizers: In traditional agriculture, farmers often fermented various organic materials, such as cow dung or plant matter, to cultivate beneficial microorganisms that could be used as biofertilizers. These natural inoculants helped improve soil health by enhancing microbial activity and nutrient availability for crops. [12] **Symbiotic Inoculation with Local Microbes**: Microbial Seeding of Agricultural Soils: In some traditional practices, farmers relied on the practice of "microbial seeding" by allowing local microorganisms to colonize soil naturally. This could be done by using local plants, mulching with plant residues, or introducing soil from other fertile areas to introduce specific beneficial microbial strains.

CHAPTER 4

DECENTRALISED PLANT BREEDING FOR NATURAL FARMING

Conventional agriculture follows centralised, institutional or commercial plant breeding for profit maximisation of seed industries. In contrast, natural farming should follow decentralised, personal or community based plant breeding for natural farming and for profit maximisation of the farmers. Plant Breeding is the most important tool in natural farming, because plant breeding can solve all the problems in agriculture. Natural farming needs customised plant breeding for each individual farmer, for their profit maximisation (Ceccarelli S, Grando S. 2006).

4.1 Objectives of Decentralised Plant Breeding:

The objectives of decentralised plant breeding for natural farming (polyculture) are completely different from the objectives of centralised plant breeding for conventional agriculture (monoculture). Here are the main objectives of decentralised plant breeding:

4.1.1 Decentralisation of Knowledge and Power:

Decentralized plant breeding has the potential to democratize knowledge and power in agriculture. Here's how: (1) **Farmer Empowerment:** Decentralized plant breeding involves farmers directly in the development of crop varieties. This enables them to tailor crops to their local conditions, increasing their agency and reducing reliance on centralized institutions or corporations. (2) **Localized Knowledge Sharing:** It promotes the exchange of traditional knowledge and practices among farmers, preserving biodiversity and adapting crops to diverse agro-ecological zones.

(3) **Reduction in Corporate Dominance**: By breaking the monopoly of large seed corporations, decentralized plant breeding shifts control over seeds back to farmers and communities, reducing dependency on patented seeds and associated costs. (4) **Community Resilience**: Decentralized approaches foster community-based seed banks and networks, enhancing resilience to climate change and market shocks. (5) **Co-creation of Innovation**: Collaborative efforts between farmers, scientists, and local organizations lead to context-specific innovations, combining modern science with traditional wisdom.

4.1.2 Adaptation to Local Conditions:

Crops are bred specifically for local agro-ecological conditions, such as soil type, climate, and pest pressures. Improved resilience to climate variability and environmental stressors.

4.1.3 Preservation of Biodiversity:

Encourages the development and conservation of diverse crop varieties, reducing genetic uniformity and vulnerability to pests and diseases. Supports the cultivation of underutilized and indigenous crops.

4.1.4 Faster Breeding Cycles:

Decentralized breeding programs are often more flexible and responsive to local needs, leading to quicker development of suitable varieties.

4.1.5 Strengthening Local Knowledge Systems:

Incorporates traditional agricultural knowledge with modern scientific methods, creating a robust knowledge-sharing ecosystem. Protects and revitalizes indigenous practices and cultural heritage.

4.1.6 Cost-Effective:

Breeding programs based on local resources and farmer collaboration are less expensive than centralized programs requiring extensive infrastructure.

4.1.7 Exclusive Varieties:

The objective of decentralised plant breeding is to develop unique, exclusive and exquisite plant varieties for each individual small-scale farmers and farming communities, for maximising the price of their products, and ultimately to maximise the profit of the farmers.

4.1.8 Premium Products:

The objective of decentralised plant breeding for natural farming is to develop unique, exclusive and exquisite quality of products, in respect of taste, flavour, colour, shape, size, texture, keeping quality, nutritive value and medicinal value, for maximising the premium price of the premium quality agricultural products.

4.1.9 Off-Season Varieties:

The objective of decentralised natural farming research is to develop off-season or all-season bearing plant varieties, to create availability of the product throughout the year, to maximise the market price of the crop and to maximise the profit of the farmers.

4.1.10 Acclimatization of Plant Varieties:

The objective of decentralised natural farming research is to develop tropical versions of the temperate crops, to create novelty in the product range and for profit maximisation of the farmers.

4.2 Methods of Decentralised Plant Breeding for Natural Farming:

The methods of decentralised plant breeding for natural farming are completely different from the methods of plant breeding followed for conventional agriculture. The traditional methods of plant breeding practiced by the indigenous people for thousands of years in different parts of the world, were studied and characterised, to develop appropriate methods of plant breeding for natural farming. These methods include:

4.2.1 Collection of Elite Plant Genotypes:

Farmers themselves collect the most profitable plant varieties of the most profitable crops for their own natural farm. The parameters of profitability will vary for each individual farmer.

4.2.2 Maintaining Plant Varieties under Multilayer Polycropping System:

Farmers should maintain their collected plant varieties (plant genetic resources) under a multilayer polycropping system, within their natural farm (multi-crop gene bank), for introgression of environmental genes and for their use in future plant breeding programme.

4.2.3 Natural Selection:

Natural selection in plant breeding refers to the process where environmental factors and ecological conditions favor the survival and reproduction of plants with advantageous traits. Unlike artificial selection, where humans control the selection process, natural selection operates without human intervention. In plant breeding, natural selection can be intentionally harnessed to develop robust and resilient crops suited to specific environments.

4.2.4 Natural Hybridization by Native Pollinator Biodiversity:

The farmer should convert their natural farming agroecosystem into a pollinator sanctuary or a pollinator hotspot by incorporating as many types of pollinator magnet plants as possible, to maximise the species richness and population of native pollinator biodiversity. This native pollinator biodiversity helps to maximise cross pollination of crops and natural hybridization of plant varieties. This open pollination system can facilitate creation of maximum plant genetic diversity.

4.2.5 Spontaneous Mutation:

Spontaneous mutation refers to changes in the DNA sequence that occur naturally without any external influence. These mutations arise due to errors in DNA replication, repair, or as a result of natural cellular processes. They play a critical role in evolution, genetic diversity, and sometimes in the development of diseases. Farmers should look for these natural events in their fields and should select the new plant types to develop new plant varieties. (Fernie AR et al. 2006).

4.2.6 Artificial Selection by the Farmers:

Artificial selection in plant breeding is a process where farmers intentionally select and breed plants with desirable traits to improve their performance, adaptability, or yield. Unlike natural selection, where environmental pressures drive evolution, artificial selection is directed by human preferences and goals.

4.2.7 Budding and Grafting:

Different types of budding and grafting techniques helps farmers to collect new plant varieties, conservation and comparison of multiple plant varieties in a single stock plant, reducing first flowering period of seed plant and maximising disease resistance and yield of crop plants.

4.2.8 In-Situ Conservation of Plant Genetic Resources (Farmers Gene Bank):

A Farmers' Gene Bank is a community-based initiative where farmers collect, conserve, and manage seeds of traditional and indigenous crop varieties. These banks empower farmers to preserve agricultural biodiversity, enhance resilience to climate change, and maintain food security by storing seeds that are well-adapted to local conditions.

4.2.9 Establishment of Vavilov Centres:

A Vavilov Centre, named after Russian scientist Nikolai Ivanovich Vavilov, refers to a geographic region identified as the origin of cultivated plants. These centers are rich in genetic diversity for particular crops and their wild relatives. Governments should establish Vavilov Centres at the centres of origin of each edible and useful plant species of the world, for conservation of plant biodiversity and inherited genetic resources.

CHAPTER 5

SEED MANAGEMENT FOR NATURAL FARMING

Seed is the only input which is used in natural farming. Traditional farmers have developed many innovative techniques for maintaining seed quality, preservation of seed and seed treatment.

Theories on Seed:

The study of seeds encompasses a variety of theories that explain their formation, development, dormancy, germination, and role in ecosystems. (1) **Evolutionary Theory of Seed Development:** Seeds evolved as an adaptive strategy to ensure the survival and dispersal of plants in diverse environments. Example: Gymnosperms and angiosperms evolved seeds to withstand harsh climates and ensure reproductive success. (2) **Double Fertilization Theory:** In flowering plants, seeds form through a unique process called double fertilization. One sperm fertilizes the egg cell to form the embryo, while another fertilizes the polar nuclei to form the endosperm. Example: The endosperm in wheat seeds is the result of double fertilization and provides energy for seedling growth. (3) **Seed Dormancy Theory:** Dormancy is an evolutionary adaptation that allows seeds to survive unfavorable conditions and germinate at the right time. (4) **Hormonal Control Theory:** Plant hormones regulate seed dormancy, germination, and development. Abscisic acid (ABA) promotes dormancy, while gibberellins (GA) stimulate germination. Ethylene, auxins, and cytokinins also play roles in seed physiology. Example: Seed priming with gibberellins promotes uniform germination in crops. (5) **Genetic Theory of Seed Traits:** Seed traits, including size, shape, and nutrient content, are determined by genetic factors. (6) **Ecological Theory of**

Seed Dispersal: Seed dispersal mechanisms evolved to maximize plant reproductive success and reduce competition. Dispersal strategies include wind (anemochory), water (hydrochory), animals (zoochory), and self-dispersal (autochory). Example: Coconut seeds are adapted for water dispersal, allowing the plant to colonize coastal areas. (7) **Physiological Seed Germination Theory**: Germination is a physiological process regulated by environmental and internal factors. Key stages include water uptake (imbibition), enzymatic activation, and radicle emergence. Optimal temperature, moisture, and oxygen are critical for successful germination. Example: Seeds of temperate crops require vernalization (cold treatment) to break dormancy. (8) Storage Behavior Theory: Seeds exhibit different storage behaviors, impacting their longevity and conservation. Orthodox seeds (such as rice) can be dried and stored at low temperatures for long periods. Recalcitrant seeds (such as mango) cannot tolerate drying and require specific conditions for storage. (9) **Energy Allocation Theory**: Seeds are evolutionary structures optimized for energy storage to support the embryo during germination and early growth. Seeds store energy as carbohydrates, proteins, or lipids, depending on the plant species. Example: Oilseeds like sunflower store energy primarily as lipids. (10) **Seed Bank Theory**: Soil seed banks play a critical role in ecosystem dynamics and plant population persistence. Seed banks consist of viable seeds stored in the soil, awaiting favorable conditions to germinate. (11) **Biochemical Theory of Seed Aging**: Aging processes in seeds involve biochemical changes that reduce viability over time. Example: Seeds stored at high humidity age faster due to increased oxidative damage. (12) Seed Priming Theory: Pre-sowing treatments (priming) enhance seed germination and vigor. Priming involves controlled hydration and re-drying to "precondition" seeds. Example: Hydropriming improves germination in drought-stressed environments.

5.1 Characteristics of Traditional Seeds:

Traditional seeds, often referred to as indigenous or heirloom seeds, have distinct characteristics that set them apart from modern hybrid or genetically modified seeds. Below are their key characteristics:

5.1.1 Genetic Diversity:

Traditional seeds exhibit a wide range of genetic variation, contributing to resilience against pests, diseases, and environmental changes. They are often well-suited to the specific agro-climatic regions where they originated.

5.1.2 Open Pollination:

Traditional seeds are usually open-pollinated, meaning they are pollinated naturally by wind, insects, or animals, allowing for seed saving and reuse by farmers.

5.1.3 Adaptation to Local Conditions:

These seeds have evolved over generations to adapt to the specific soil, climate, and ecological conditions of their region, making them reliable under traditional farming systems.

5.1.4 Can Grow Without External Inputs:

Traditional seeds often thrive without the need for chemical fertilizers, pesticides, or intensive irrigation, aligning with natural farming practices.

5.1.5 Resilience to Stress:

They are more resilient to environmental stress, such as droughts, floods, and temperature extremes, compared to some hybrid varieties.

5.1.6 Nutritional Value:

Crops grown from traditional seeds often have higher nutritional content, richer flavors, and better culinary qualities than those from hybrid or GMO seeds.

5.1.7 Self-Renewing:

Farmers can save seeds from their harvests to plant in subsequent seasons, promoting self-reliance and reducing dependency on commercial seed companies.

5.1.8 Cultural Significance:

Traditional seeds are often tied to local customs, festivals, and food traditions, preserving the agricultural heritage and identity of communities.

5.1.9 Pest and Disease Resistance:

Many traditional seeds have natural resistance to certain pests and diseases due to their genetic diversity and long adaptation period.

5.1.10 Sustainability:

They support sustainable farming practices by maintaining biodiversity and reducing the need for external inputs like synthetic chemicals.

5.1.11 Longevity:

Traditional seeds remain in use for very long periods of time, often hundreds of years, due to its wide genetic pools and wide adaptability, whereas modern hybrid seeds usually last for less than 10 years in the field, requiring quick replacement of varieties.

5.1.12 Lower Yields (Comparatively):

While traditional seeds are highly sustainable, they often produce lower yields compared to modern hybrids, making them less appealing for large-scale commercial farming.

5.1.13 Lower Risk:

While traditional seeds are comparatively low fielders, the risks of crop loss due to extreme weather events or due to incidence of pests or diseases is very less.

5.2 Objectives Seed Quality Management:

The management of the quality of traditional seeds is essential to preserve their integrity, ensure sustainable agriculture, and support farmers' livelihoods. Below are the key objectives of quality management for traditional seeds:

5.2.1 Preservation of Genetic Diversity:

To maintain the genetic integrity and diversity of traditional seed varieties, ensuring they remain viable for future generations. Protecting indigenous seed varieties from genetic erosion caused by hybridization or cross-pollination with commercial seeds.

5.2.2 Enhancement of Seed Viability and Germination Rates:

Ensuring that seeds retain their ability to germinate and produce healthy plants through proper storage, treatment, and testing methods.

5.2.3 Adaptation to Local Conditions:

To maintain and enhance the adaptability of traditional seeds to specific agro-climatic conditions, ensuring resilience against local environmental stresses like drought, pests, and diseases.

5.2.4 Sustainability in Farming Practices:

Supporting organic and natural farming systems by providing high-quality traditional seeds that do not require chemical inputs.

5.2.5 Preservation of Nutritional and Culinary Value:

Ensuring the continuation of traditional crops that are known for their superior nutritional content and unique flavors, contributing to local food security and dietary diversity.

5.2.6 Minimization of Contamination Risks:

Protecting traditional seeds from contamination by genetically modified organisms (GMOs), diseases, or pests during cultivation, storage, or transportation.

5.2.7 Promoting Seed Saving Practices:

Facilitating practices that allow farmers to save and reuse seeds without compromising quality, thereby reducing dependency on external seed suppliers.

5.2.8 Improvement of Storage Techniques:

Developing and implementing proper seed storage methods to extend shelf life, prevent spoilage, and maintain genetic integrity.

5.2.9 Support for Sustainable Livelihoods:

Ensuring that high-quality traditional seeds are available to farmers, empowering them to cultivate crops that are economically viable and culturally significant.

5.2.10 Compliance with Certification Standards:

Ensuring traditional seeds meet national and international quality standards, such as purity, germination percentage, and health, for improved marketability.

5.2.11 Promotion of Agro-Biodiversity Conservation:

Encouraging the cultivation and preservation of a wide variety of traditional crops to support ecological balance and reduce dependency on monoculture systems.

5.2.12 Education and Awareness:

Spreading knowledge among farmers about the importance of seed quality management and the techniques for maintaining and improving traditional seed stocks.

5.2.13 Protection of Indigenous Knowledge:

Documenting and preserving traditional practices related to seed management, ensuring that indigenous knowledge systems are recognized and utilized effectively.

5.2.14 Encouragement of Community Participation:

Promoting community seed banks, farmer cooperatives, and local initiatives to enhance the collective management of traditional seed resources.

5.3 Traditional Methods of Seed Preservation:

Traditional methods of seed preservation are deeply rooted in indigenous knowledge systems and have been practiced for generations to ensure the availability of high-quality seeds for future planting. These methods focus on natural, sustainable, and cost-effective techniques. Here are some key traditional methods of seed preservation:

5.3.1 Sun Drying:

Seeds are dried under sunlight to reduce moisture content, which helps in preventing fungal growth and prolongs their shelf life. Care is taken to avoid overexposure to prevent damage to the seeds.

5.3.2 Clay Pot Storage:

Seeds are stored in airtight clay pots, often with natural desiccants like ash, lime, or charcoal, to absorb moisture and prevent pest infestations.

5.3.3 Ash Coating:

Seeds are mixed with wood ash, which acts as a natural pesticide and helps keep the seeds dry. This method is particularly effective for storing cereals and legumes.

5.3.4 Cow Dung Coating:

Seeds are coated with a paste of cow dung and ash, then dried. This protective layer prevents pest attacks and fungal infections.

5.3.5 Neem Leaf Layering:

Seeds are layered with neem leaves, which have natural antifungal and insect-repellent properties, to protect them from pests.

5.3.6 Storing in Gourds or Bamboo Containers:

Hollowed gourds or bamboo containers are used as natural storage vessels, which are often sealed with clay or mud to keep out air and moisture.

5.3.7 Underground Storage:

Seeds are stored in pits or underground granaries lined with dry straw or ash to protect them from extreme temperatures and pests.

5.3.8 Mud Plastering:

Storage containers made of mud or earthen materials are plastered to create a sealed environment, keeping out pests and maintaining a stable temperature.

5.3.9 Use of Natural Oils:

Seeds are treated with natural oils such as mustard oil, castor oil, or neem oil to prevent fungal growth and pest infestations.

5.3.10 Salt Mixing:

Seeds are mixed with salt, which acts as a desiccant and repels pests, particularly in humid climates.

5.3.11 Smoke Storage:

Seeds are stored in spaces where smoke from wood fires circulates, such as thatched roofs or storage huts. The smoke acts as a natural preservative.

5.3.12 Husk or Pod Storage:

Certain seeds, like groundnuts or grains, are stored in their natural husks or pods to provide an extra layer of protection.

5.3.13 Community Seed Banks:

Farmers collectively store seeds in community seed banks, where seeds are preserved using traditional methods and exchanged as needed.

5.3.14 Use of Sand Layers:

Seeds are layered with dry sand in storage containers, which helps absorb moisture and protect seeds from insects.

5.3.15 Charcoal Layering:

Charcoal is used in storage containers as a desiccant and pest repellent to maintain seed quality.

5.3.16 Hanging Storage:

Seeds are stored in bundles or bags and hung from ceilings or rafters to protect them from rodents and moisture.

5.4 Traditional Methods of Seed Treatment:

Traditional methods of seed treatment before sowing are essential for enhancing germination rates, protecting seeds from pests and diseases, and ensuring healthy crop growth. These methods rely on natural and readily available materials, making them eco-friendly and cost-effective. Here are some widely used traditional seed treatment techniques:

5.4.1 Cow Dung and Cow Urine Treatment:

Seeds are soaked in a mixture of cow dung and cow urine, which acts as a natural disinfectant and nutrient booster. This method is particularly effective in enhancing seed vigor and protecting against soil-borne pathogens.

5.4.2 Ash Coating:

Seeds are coated with fine wood ash, which acts as an antifungal agent and protects them from pests. This method is commonly used for grains like millet and sorghum.

5.4.3 Neem Leaf Extract:

Seeds are soaked in a neem leaf decoction or coated with neem leaf powder. Neem's natural antifungal and insect-repellent properties protect seeds from fungal infections and pests.

5.4.4 Turmeric Powder Coating:

Seeds are mixed with turmeric powder, which has antimicrobial properties, to prevent fungal infections and diseases. This is especially effective for pulses and legumes.

5.4.5 Salt Water Treatment:

Seeds are soaked in saltwater to remove damaged or hollow seeds that float. This ensures that only healthy seeds are selected for sowing.

5.4.6 Cow Butter and Lime Treatment:

Seeds are coated with a mixture of cow butter and lime to prevent fungal infections and improve germination rates. This method is traditional in some regions for rice and wheat seeds.

5.4.7 Clay and Dung Slurry:

Seeds are dipped in a slurry made of clay and cow dung, forming a protective layer that prevents pest attacks and improves seed-to-soil contact.

5.4.8 Hot Water Treatment:

Seeds are soaked briefly in warm water to kill surface pathogens and pests. Care is taken to maintain the right temperature to avoid damaging the seeds.

5.4.9 Soaking in Herbal Extracts:

Seeds are soaked in extracts of locally available herbs like tulsi, garlic, or onion, which have antifungal and antibacterial properties.

5.4.10 Castor Oil Coating:

Seeds are treated with castor oil to protect them from soil-borne pests like termites and ants. This method is widely used for groundnut seeds.

5.4.11 Milk Soaking:

Seeds are soaked in cow or goat milk, which acts as a natural fungicide and nutrient source. This is often done for small vegetable seeds.

5.4.12 Tamarind Pulp Treatment:

Seeds are soaked in tamarind pulp extract to remove seed coats or enhance germination, particularly for hard-coated seeds like tamarind and legumes.

5.4.13 Charcoal Powder Coating:

Seeds are dusted with charcoal powder to absorb moisture and protect them from fungal growth. This method also ensures better seed storage before sowing.

5.4.14 Fermentation Process:

Seeds are soaked in water and left to ferment for a short period. This treatment softens hard seed coats and promotes better germination.

5.4.15 Urine and Water Soaking:

Human or animal urine diluted with water is used to soak seeds for a short duration. This method provides nitrogen to the seeds and acts as a natural disinfectant.

5.4.16 Lemon or Tamarind Juice:

Acidic juices like lemon or tamarind are used to scarify hard-coated seeds, making them easier to germinate.

5.4.17 Aloe Vera Gel Coating:

Seeds are coated with aloe vera gel, which has antifungal and antibacterial properties and helps retain moisture.

5.4.18 Seed Scarification:

Seeds with hard coats are manually scratched or rubbed against a rough surface to break dormancy and enhance germination.

5.4.19 Herbal Smoke Treatment:

Seeds are exposed to smoke from burning medicinal plants like neem or tulsi to disinfect and prepare them for sowing.

5.4.20 Mud and Ash Mixture:

Seeds are coated with a paste of mud and ash, which helps prevent pest attacks and improves seed-to-soil contact during sowing.

5.5 Traditional Knowledge about Seed:

Traditional knowledge and practices about seeds are an integral part of indigenous and agricultural communities, reflecting centuries of experience, experimentation, and adaptation.

5.5.1 Seed Selection:

Farmers select seeds from the healthiest, most productive, and disease-resistant plants. Ensures improved crop varieties and adaptation to local conditions over generations.

5.5.2 Seed Preservation and Storage:

Use of natural materials and methods to store seeds and protect them from pests, moisture, and spoilage. Storing seeds in clay pots, gourds, or bamboo containers. Using neem leaves, ash, or turmeric to repel pests. Smoking seeds to reduce moisture and increase shelf life.

5.5.3. Seed Exchange Systems:

Barter or exchange of seeds among farmers within communities or across regions. Promotes genetic diversity and resilience in farming systems. Example: Traditional seed fairs or "seed banks" where farmers share seeds and knowledge.

5.5.4 Seed Sovereignty:

Farmers hold the right to save, reuse, and share seeds without external dependency. Ensures independence from commercial seed systems and protects indigenous varieties.

5.5.5 Landrace Cultivation:

Cultivating traditional seed varieties (landraces) that are well-suited to local climates and soils. Greater resilience to pests, diseases, and climate variability. Enhanced taste, nutrition, and cultural significance. Example: Indigenous rice varieties in India such as "Kala Namak" and "Gobindobhog."

5.5.6 Germination Testing:

Testing seed viability before planting by observing their sprouting ability. Seeds are soaked in water or placed on wet cloth to monitor germination. Example: Sprouting seeds of lentils or pulses before sowing.

5.5.7 Rituals and Cultural Practices:

Seeds are often associated with rituals, prayers, and ceremonies to invoke blessings for a good harvest. In many Indian communities, seeds are blessed during festivals like Makar Sankranti. Special seed varieties are preserved as sacred and used in specific cultural events.

5.5.8 Crop Rotation and Mixed Cropping:

Rotating crops or intercropping to preserve soil fertility and enhance seed quality. Reduces the build-up of pests and diseases, ensuring robust seeds.

5.5.9 Seed Knowledge Transmission:

Oral transfer of seed-related knowledge from elders to younger generations. Preserves traditional practices and ensures their continuity.

5.5.10 Community Seed Banks:

Collective storage of seeds to preserve traditional varieties and support local farmers. Example: Farmers in South India maintain "seed huts" to safeguard indigenous millet and rice varieties.

5.5.11 Natural Seed Treatments:

Treating seeds with natural substances to improve germination and resistance to pests and diseases. Coating seeds with cow dung or buttermilk. Using wood ash or lime as a seed dressing.

5.5.12 Seasonal Knowledge:

Timing seed sowing with lunar cycles, weather patterns, or specific seasons. Ensures optimal germination and growth.

5.5.13 Conservation of Wild Varieties:

Preserving and cultivating wild relatives of crops as a reservoir of genetic diversity. Provides a buffer against pests, diseases, and climate change.

5.5.14 Seed Festivals and Celebrations:

Organizing community events to celebrate the importance of seeds and share knowledge. "Beej Parab" (Seed Festival) in tribal communities of India. Traditional ceremonies in Africa to honor seeds before sowing.

5.5.15 Focus on Local Adaptation:

Seeds are selected and adapted to local environmental conditions over generations. Ensures that seeds thrive in the specific climate, soil, and pest conditions of the region.

CHAPTER 6

CROPPING SYSTEM INNOVATION FOR NATURAL FARMING

6.1 Theories on Plant Assemblage:

Plant assemblage refers to the grouping or community of plant species in a particular area due to ecological interactions, environmental factors, and evolutionary processes. Several theories explain how plant assemblages form and persist. These theories are broadly categorized into ecological, evolutionary, and biogeographical perspectives:

6.1.1 Ecological Theories:

These theories focus on the role of environmental conditions and interactions between species in determining plant assemblages.

a. **Niche Theory:** Plants coexist because they occupy different niches, defined by specific environmental factors such as light, temperature, soil type, and moisture. Resource partitioning reduces competition and allows diverse species to coexist.

b. **Neutral Theory:** Proposes that all species are ecologically equivalent and that randomness (e.g., dispersal, extinction) governs plant assemblages. Emphasizes the role of stochastic processes over deterministic factors.

c. **Succession Theory:** Explains plant assemblages as a dynamic process where species composition changes over time due to environmental modifications. **Primary Succession:** Occurs in newly formed areas (e.g., lava flows, glacial retreats). **Secondary Succession:** Happens in areas disturbed but with existing soil (e.g., after forest fires).

d. **Intermediate Disturbance Hypothesis:** Suggests that plant diversity is highest at intermediate levels of disturbance. Frequent disturbances prevent competitive exclusion, while rare disturbances allow dominant species to monopolize resources.

e. **Competition and Facilitation Theory: Competition:** Plants compete for limited resources such as water, nutrients, and light. **Facilitation:** Some plants improve conditions for others, such as nitrogen-fixing legumes enriching soil.

6.1.2 Evolutionary Theories:

These theories emphasize historical and evolutionary processes in shaping plant assemblages.

a. **Coevolutionary Theory:** Plant assemblages evolve in response to mutualistic and antagonistic interactions, such as pollination and herbivory. Adaptations lead to complementary assemblages of species.

b. **Phylogenetic Clustering:** Closely related species often occur together because they share similar ecological requirements and evolutionary history.

c. **Adaptive Radiation:** A single ancestral species diversifies into multiple species adapted to different ecological niches, contributing to species-rich assemblages.

6.1.3 Biogeographical Theories:

These theories consider the spatial distribution of plants and historical events like continental drift.

a. **Island Biogeography Theory:** Plant assemblages on islands depend on immigration, extinction rates, and island characteristics (size, isolation). Larger, closer islands host more diverse assemblages.

b. **Species–Area Relationship**: Larger areas tend to support more plant species due to increased habitat heterogeneity.

c. **Historical Contingency**: Historical events (e.g., glaciation, volcanic eruptions) shape the current composition of plant assemblages.

6.1.4 Modern Integrative Approaches:

Metacommunity Theory: Explains assemblages as a balance between local processes (e.g., competition) and regional dynamics (e.g., dispersal).

Trait-Based Ecology: Focuses on the functional traits of plants, such as leaf area or root depth, and their role in assemblages.

Ecosystem Engineering: Examines how plants modify their environment, creating conditions for other species.

6.2 Theories on cropping System:

Theories related to cropping systems provide insights into how crops are grown, managed, and rotated to optimize productivity, sustainability, and environmental health. These theories combine principles of agronomy, ecology, economics, and sociology to explain and improve agricultural practices.

6.2.1 Crop Rotation Theory:

Rotating crops improves soil fertility, reduces pest and disease buildup, and enhances productivity.

6.2.2 Multiple Cropping Theory:

Growing multiple crops in the same field (sequentially or simultaneously) maximizes land use efficiency and diversifies income. Intercropping, relay cropping, and mixed cropping are examples of multiple cropping. Complementary resource use (e.g., light, nutrients) between crops boosts overall productivity. Diversifying crops improves resilience to market and climatic risks. Example: Maize and beans intercropped to maximize yield and maintain soil fertility. (3)

6.2.3 Agroecological Theory:

Cropping systems should mimic natural ecosystems to achieve sustainability. Biodiversity, nutrient cycling, and ecological balance are key to sustainable farming.

6.2.4 Monoculture vs. Polyculture Theory:

Polyculture systems (growing multiple crops) are more resilient than monocultures (single-crop systems). Monocultures are efficient but vulnerable to pests, diseases, and market fluctuations. Polycultures enhance biodiversity, soil health, and ecosystem stability. Example: A polyculture of maize, beans, and squash (the "Three Sisters" system) ensures food security and soil health.

6.2.5 Integrated Farming Systems Theory:

Integrating crops with livestock, aquaculture, or forestry creates a holistic, resource-efficient farming system. By-products of one component (e.g., livestock manure) are inputs for another (e.g., crop fertilization). Diversification reduces risks and enhances resource use efficiency.

6.2.6 Resource Use Efficiency Theory:

Cropping systems should maximize the efficient use of resources like water, nutrients, and light. Crops with different growth habits or rooting depths utilize resources more efficiently. Efficient resource use reduces costs and environmental impacts.

6.2.7 Crop Diversification Theory:

Diversifying crops in a farming system enhances resilience to climatic and market risks.

6.2.8 Continuous vs. Fallow Cropping Theory:

Continuous cropping maximizes land use, while fallow periods restore soil fertility and moisture. Example: Shifting cultivation incorporates fallow phases for soil restoration.

6.2.9 Economic Theory of Cropping Systems:

Cropping systems should maximize economic returns while considering environmental and social sustainability. Market demand, input costs, and risk factors influence crop choices. Diversification and value addition enhance profitability. Cropping systems must balance profitability with long-term sustainability.

6.2.10 Sustainability Theory in Cropping Systems:

Sustainable cropping systems meet current needs without compromising future generations' ability to meet theirs. Example: Using no-till farming and mulching to conserve soil structure and moisture.

6.2.11 Climate Resilience Theory:

Cropping systems should adapt to changing climatic conditions to ensure productivity and stability. Drought-resistant, flood-tolerant, and short-duration crop varieties are critical. Agroforestry and mixed cropping reduce vulnerability to extreme weather events.

6.2.12 Nutrient Cycling Theory:

Efficient nutrient cycling within cropping systems enhances soil fertility and reduces external inputs. Diverse cropping systems support nutrient cycling and reduce leaching losses.

6.2.13 Ecological Intensification Theory:

Enhancing ecological processes within cropping systems improves productivity and sustainability. Pollination, pest control, and soil health are improved by biodiversity and ecosystem services. Ecological intensification supports long-term productivity and environmental health.

6.2.14 Food Security and Cropping System Theory:

Cropping systems must be designed to ensure food security for a growing population. High-yield systems, diversification, and efficient resource use are critical. Sustainable intensification balances productivity and environmental health.

6.3 Objectives of Cropping System Innovation for Natural Farming:

Natural farming provides infinite opportunities to the farmers, to innovate unique cropping systems for themselves. Individual farmers can design their own unique cropping system for minimising the risk and cost of production of crops and to maximise crop diversity, aggregate crop yield and profitability of farming. There is absolutely no limit to the number of crops (intercrops) that can be incorporated in the mixed cropping system and the numbers of layers of crops can be accommodated in the crop assemblage (crop community). Here are the main objectives of the cropping system Innovation in natural farming.

6.3.1 Development of Appropriate Multilayer Polycropping System for Each Individual Farmer:

Natural farming agroecosystems are artificial (man-made) ecosystems which try to mimic the structure and function of natural forest ecosystems, to automatically maximise their ecosystem services of crop production and crop protection. Each individual farmer should develop their own unique multilayer polycropping system, by arranging their plant biodiversity in a multicrop, multilayer structure, so that different plant species and plant varieties complement each other to maximise their ecosystem services (Complementarity Theory).

6.3.2 Conservation of Natural Capital:

Natural capital is the natural resources available in an agricultural farm. Natural resources include soil, water, solar energy, temperature, relative humidity, rainfall, wind velocity, seasonal variation in climatic conditions and biodiversity. One of the main aim of designing the cropping system of each individual farm is to conserve the natural resources, to make the farm a self-sustaining unit of agriculture, through maximising water harvesting, carbon sequestration and nitrogen sequestration.

6.3.3 Conservation of biodiversity:

Natural farming (polyculture or ecological agriculture) is a biodiversity based agriculture, because it solely depends on native biodiversity of organisms such as plants, animals and microorganisms, for optimising crop production and crop protection. Natural farming adopts various traditional practices for conservation of biodiversity in its agroecosystem such as water harvesting, no-till, no off-farm input, mulching, cover cropping, multilayer polycropping, mixed animal husbandry, beekeeping and mixed aquaculture. Multilayer polycropping system is the best cropping system invented so far, by the farmers for conservation of biodiversity in soil and above ground.

6.3.4 Development of nature based solutions (NbS):

Nature based solutions (NbS) are those solutions, which uses living organisms such as plants, animals and microorganisms (biology), instead of using chemicals (chemistry) or machines (physics) as their primary tools, to solve the problems of mankind. Cropping systems innovations in natural farming such as multilayer polycropping systems can provide solutions to many global problems such as climate change, biodiversity loss, groundwater depletion, low rainfall, low soil fertility, hunger, malnutrition, poverty, unemployment, diseases and depression.

6.3.5 Adaptation and Mitigation of climate change:

The cropping systems of natural farming agroecosystem (multilayer polycropping) can help carbon sequestration in soil, plant, animals and microorganisms and can reduce the emission of greenhouse gases such as CO_2, methane and nitrous oxide. The compact permanent vegetation of natural farming agroecosystems can buffer the atmospheric temperature and soil temperature within the agroecosystem, increase relative humidity through transpiration, can cause cloud seeding through biogenic volatile organic compounds (BVOC) and ice nucleating bacteria and can induce rainfall. All these functions of the plant biodiversity in natural farming agroecosystems help adaptation and mitigation of climate change.

6.3.6 Automation of Ecosystem Services:

The multilayer polycropping system in a natural farm resembles a natural forest ecosystem in respect of both its structure and function and provides all the ecosystem services of a natural forest such as water harvesting and conservation, biomass synthesis, carbon sequestration, nitrogen sequestration, conservation of biodiversity, mitigation of climate change and natural control of pests and diseases. The most important ecosystem service is the provisioning ecosystem service for food, water, medicine, and other useful materials.

6.3.7 Sustainability and Resilience:

Sustainability in Polyculture: (1) **Biodiversity Conservation**: Polyculture increases species diversity, creating habitats for beneficial organisms like pollinators and predators. (2) **Improved Soil Health**: Diverse root systems improve soil structure and nutrient cycling. Nitrogen-fixing crops (e.g., legumes) replenish soil fertility, reducing the need for synthetic fertilizers. Organic matter from crop residues enhances soil microbial activity. (3) **Efficient Resource Use**: Diverse crops use sunlight, water, and nutrients more efficiently. Minimizes resource wastage compared to monoculture systems. (4) **Zero Chemical Use**: Mixed crops reduce pest and disease pressure, minimizing the need for pesticides. Eliminates environmental contamination and promotes safe food production. (5) **Climate Change Mitigation**: Polyculture systems store more carbon in biomass and soil than monocultures. Reduces greenhouse gas emissions by avoiding chemical use.

Resilience in Polyculture: [1] **Pest and Disease Control**: Diverse plant species disrupt pest cycles, making outbreaks less severe. Reduces the risk of total crop failure due to pest infestations. (2) **Adaptation to Climate Variability**: Different crops respond differently to weather extremes, ensuring some yield even in adverse conditions. Reduces dependence on a single crop vulnerable to specific climate risks. (3) **Economic Stability**: Growing multiple crops diversifies income sources, reducing financial risks. Provides flexibility in market dynamics as farmers can sell different products. (4) **Ecosystem Services**: Promotes natural pollination, water retention, and erosion control. Strengthens ecological balance, reducing susceptibility to environmental shocks. (5) **Long-Term Productivity**: Polyculture systems avoid the resource depletion common in monocultures. Encourages regeneration of soil and ecosystem functions over time.

6.3.8 Profit Maximisation for the Farmers:

The main objective of cropping system innovation in natural farming is profit maximisation of the farmers. Multiple factors contribute to the profit maximisation game. (1) **Minimise Risk:** The probability of risk of crop loss depends on the number of crops present in the multilayer polycropping system of natural farming agroecosystems. (2) **Minimise Cost:** Plant biodiversity provides organic biomass which forms natural mulch on soil surface. Plant biodiversity supports microorganisms to decompose organic matter and mineralise plant nutrients, to fix atmospheric nitrogen and to solubilise phosphate, to provide plant nutrition free-of-cost. Plant biodiversity host natural enemies and animal biodiversity to control pests, diseases and weeds, free-of-cost. (3) **Maximise Aggregate Yield:** Highest possible cropping intensity in a multilayer polycropping system provides the highest aggregate yield of crops. (4) **Maximise Quality of Products:** Avoidance of chemicals and optimisation of plant nutrition and crop protection maximise the quality of the agricultural products. (5) **Maximise Product Price:** Due to diversified product range and high quality of products, natural farmers have the liberty to fix the price of their own products, and directly sell the products to the consumers at a premium price. (6) **Regular Income:** Different types of natural farming products are harvested on a daily basis, which provides a regular passive income to the farmers. All these features of the natural farming cropping system can open up multiple sources of income and can maximise the profit of the farmer.

6.4 Methods of Cropping System Innovation for Natural Farming:

Cropping system innovations are very common in traditional agricultural landscapes of the world however it is conspicuously absent in conventional agriculture (monoculture). Here are some techniques that might be helpful for the modern farmers to develop new cropping systems, for their own unique models of natural farming.

6.4.1 Agroecosystem Analysis of Traditional Agricultural Landscape:

Agroecosystem analysis (AEA) of traditional agricultural landscapes involves assessing their ecological, social, and economic components to understand their sustainability and functionality. Below are the key methods used for analyzing such systems: (1) **Ecological Analysis:** (a) **Biodiversity Assessment: Species Inventory:** Document plant, animal, and microbial species in the system. **Crop Diversity Studies:** Record traditional cropping patterns, polycultures, and crop rotations. **Indicator Species:** Identify species that signify ecosystem health or degradation. (b) **Soil Analysis: Soil Sampling:** Evaluate organic matter, nutrient content, pH, and texture. **Traditional Soil Knowledge:** Incorporate local methods of assessing soil fertility, such as color, texture, or water retention. (c) **Water Resource Evaluation: Hydrological Mapping:** Identify traditional water sources like ponds, streams, or rainwater harvesting structures. **Irrigation Practices:** Analyze traditional water management techniques and their efficiency. (d) **Climate Resilience Assessment:** Assess how traditional practices mitigate climate variability (e.g., drought-resistant crops, soil conservation methods). (2) **Agronomic Analysis:** (a) **Cropping Systems Analysis: Cropping Patterns:** Study traditional monocropping, mixed cropping, and intercropping systems. **Productivity Metrics:** Evaluate yields and input-output ratios. (b) **Pest and Disease Management:** Document traditional pest control methods, such as natural repellents or companion planting. Assess the role of agroecological diversity in pest suppression. (c) **Nutrient Cycling:** Study traditional practices of nutrient recycling, such as composting, green manuring, or animal integration. (3) **Socio-Cultural**

Analysis: (a) **Knowledge Systems:** Document local agricultural knowledge, practices, and decision-making processes. Analyze how knowledge is transmitted across generations. (b) **Labor and Gender Roles:** Examine labor division, gender roles, and their impact on agricultural productivity. (c) **Cultural Significance:** Investigate the role of agriculture in festivals, rituals, and traditional governance. (4) **Economic Analysis:** (a) **Cost–Benefit Analysis:** Compare inputs (labor, seeds, tools) and outputs (yields, profits) to measure economic sustainability. (b) **Market Linkages:** Assess the availability of local markets for traditional crops and the impact of modern market forces. (c) **Resource Use Efficiency:** Evaluate the efficiency of land, water, and labor use in traditional systems. (5) **Landscape and Spatial Analysis:** (a) **Land Use Mapping:** Use GIS and remote sensing to map traditional land-use patterns, crop zones, and agroforestry systems. (b) **Microclimate Analysis:** Study the role of landscape features (e.g., trees, hedgerows) in modifying local climate conditions. (c) **Connectivity Studies:** Assess the connectivity between agricultural, forest, and grazing systems in traditional landscapes. (6) **Participatory Rural Appraisal (PRA):** Engage with local farmers through focus group discussions, transect walks, and participatory mapping to capture indigenous insights. (7) **Comparative Analysis:** Compare traditional agricultural landscapes with modern systems in terms of sustainability, resilience, and productivity. (8) **Sustainability Metrics:** Apply frameworks like the Sustainable Livelihoods Approach (SLA) or FAO's Sustainability Assessment of Food and Agriculture Systems (SAFA) to evaluate environmental, economic, and social sustainability. **Challenges:** Loss of traditional knowledge due to modernization. Lack of accurate records or standardized metrics. Difficulty in integrating traditional practices with modern scientific analysis.

6.4.2 Characterisation of Traditional Agricultural Knowledge:

Characterization of traditional agricultural knowledge (TAK) involves identifying, documenting, and analyzing the unique practices, techniques, and cultural beliefs associated with traditional farming systems. Here are some common methods: (1) **Documentation and Archiving**: (a) **Interviews and Oral Histories**: Conduct interviews with elder farmers, community leaders, and practitioners to capture their agricultural practices and cultural significance. (b) **Ethnographic Studies**: Detailed documentation of rituals, festivals, and local practices linked to agriculture. (c) **Audio–Visual Records**: Use photographs, videos, and audio recordings to preserve practices. (d) **Literature Review**: Analyze historical texts, folklore, and local manuscripts for agricultural knowledge. (2) **Participatory Rural Appraisal (PRA)**: Involves engaging local communities to map their resources, seasonal calendars, and farming practices. Techniques include focus group discussions, transect walks, and participatory mapping. (3) **Scientific Validation**: (a) **Experimental Studies**: Test traditional practices (e.g., seed treatment, pest control methods) in controlled environments to validate their efficacy. (b) **Comparative Studies**: Compare traditional methods with modern agricultural practices to highlight their advantages and limitations. (4) **Classification Systems**: Categorize TAK into domains like soil fertility, water management, pest control, crop diversity, and post-harvest practices. Create taxonomies for plant species, pest management strategies, and cropping systems. (5) **Geo–Spatial Mapping**: Use GIS and remote sensing to map traditional land-use patterns, cropping systems, and water management techniques. (6) **Socio–Cultural Analysis**: Study the link between agricultural practices and cultural traditions, religious beliefs, and community structures. Analyze gender roles and knowledge transfer within families and communities. (7) **Policy and Legal Analysis**: Examine how traditional knowledge is recognized and protected under laws like the Biological Diversity Act and farmers' rights frameworks. (8) **Digital Platforms and Databases:** Develop online repositories for TAK, such as databases for traditional seeds, soil management techniques, and pest control methods. Platforms like the Traditional Knowledge Digital Library (TKDL) can serve as models. (9) **Integration with Modern Science:** Collaboration between local farmers and scientists to co-develop sustainable farming practices. Document hybrid techniques that incorporate TAK with modern agricultural innovations. **Challenges in Characterization:** Loss of knowledge due to urbanization and generational shifts. Language and cultural barriers in documentation. Ethical concerns in ensuring fair recognition and benefits to knowledge holders.

6.4.3 Inventory of Crop and Livestock Biodiversity:

International organizations or national governments should publish an open access online database of all crop varieties and livestock breeds available in the world. This database should include a rating and ranking of the plant varieties and animal breeds, on the basis of their nutritive value, medicinal value, economic value and ecological value. AI based chatbots should provide customised portfolio suggestions of crop varieties and livestock breeds to each individual farmer, by searching from this database. For example the **Global Inventory of Floras and Traits (GIFT)** holds over 3,800 species lists, covering 80% of known plant species globally, aiding in biodiversity analysis and conservation efforts (Weigelt P et al. 2019). Profitability analysis of crops also started (Parcell J, Cain W. 2014).

6.4.4 Plant and Animal Portfolio Optimisation for Natural Farming:

Plant and animal portfolio optimization in natural farming involves the strategic selection, integration, and management of both plant and animal species to create a balanced, sustainable, and mutually beneficial farming system. This approach aims to optimize the productivity, biodiversity, and ecological health of the farm by leveraging the interactions between plants, animals, and the environment. By mimicking natural ecosystems, this method promotes resilience, reduces inputs, and enhances long-term sustainability. (1) **Biodiversity and Ecosystem Services:** The inclusion of diverse plant and animal species ensures that the farm ecosystem functions similarly to a natural habitat. A variety of species can provide important ecosystem services such as pest control, pollination, soil fertility, and water regulation. Animals can help control pests and weeds, while plants can provide shelter, food, and other resources for beneficial insects and wildlife. Together, they create a more resilient and self-sustaining farming system. (2) **Synergy Between Plants and Animals:** (a) **Nutrient Cycling:** Animals, particularly livestock, contribute to nutrient cycling through their manure, which can be used to fertilize plants. In turn, plants can offer fodder for animals, creating a closed loop that reduces the need for external inputs. (b) **Weed and Pest Management:** Livestock, such as chickens or goats, can help manage weeds and pests naturally. Their grazing habits can reduce the need for herbicides and pesticides while improving soil health through the addition of organic matter. (5) **Closed-Loop Systems:** (a) **Waste Recycling:** Both plant and animal waste can be recycled within the system to minimize external inputs. Animal manure can be composted and used as fertilizer for crops, while plant residues can be used to feed livestock or provide mulch for the soil. This reduces dependency on synthetic fertilizers, herbicides, and pesticides.

6.4.5 Farm Landscape Design for Natural Farming:

The design and layout of farm landscapes is very important for natural farming agroecosystems.

Objectives of Farm Landscape Design:

The objectives of farm landscape design are to optimize productivity, sustainability, and ecosystem services by mimicking natural ecosystems. These objectives balance environmental health, economic viability, and social well-being. Key objectives of farm landscape design: (1) **Maximizing Productivity**: To ensure year-round production of diverse crops and resources. Design planting zones with complementary crops to maximize yields. Use vertical layering (e.g., trees, shrubs, ground cover) to optimize sunlight utilization. Incorporate multi-purpose plants for food, fodder, and other uses. (2) **Enhancing Biodiversity**: To foster ecological balance and resilience. Plant a mix of native and adaptive species that attract pollinators and natural pest predators. Create habitats for wildlife, such as ponds, hedgerows, and shelterbelts. Avoid monoculture patches; integrate diverse crops and wild areas. (3) **Soil Health Improvement**: To build soil health and soil structure for long-term sustainability. Incorporate nitrogen-fixing plants (e.g., legumes). Use cover crops and mulching to prevent erosion and improve organic matter. (4) **Water Management**: To harvest, conserve, and use water efficiently. Design contour-based swales and ponds to capture rainwater. Grow drought-resistant and water-intensive crops in separate zones based on water availability. (5) **Energy Efficiency**: To reduce reliance on external energy inputs and maximize renewable energy use. Plant windbreaks to reduce energy loss from wind exposure. Practice no-till farming. (6) **Climate Resilience**: To adapt to and mitigate the effects of climate change. Diversify crops to spread risks associated with weather variability. Use agroforestry to buffer against temperature extremes and sequester carbon. Protect the farm with natural barriers against floods or wind. (7) **Economic Viability**: To ensure steady income from diverse outputs. Integrate cash crops with subsistence crops for a balanced economy. Include livestock, aquaculture, or agroforestry for diversified income streams. Create value-added products (e.g., jams, herbal teas) to enhance marketability. (8) **Waste Reduction and Recycling**: To close nutrient and energy loops. Compost farm waste and use it as fertilizer. Use greywater systems for irrigation. Utilize livestock manure in nutrient cycling. (9) **Aesthetic and Recreational Value**: To create a visually appealing and peaceful environment. Include flowering plants and aesthetic landscaping in design. Create walking paths, seating areas, or educational zones for visitors. (10) **Community Engagement and Education**: To share knowledge and involve local communities. Host workshops, farm tours, or farmers' markets. Engage neighbors in collaborative farming or resource sharing. Demonstrate sustainable practices to encourage adoption. **Example: Permaculture Design**: (1) **Zones and Layers**: Zone 1: Intensive vegetable garden near the farmhouse for daily use. Zone 2: Orchards with mixed fruit and nut trees. Zone 3: Pastures for rotational grazing or staple crops. Vertical layers: Canopy trees, understory shrubs, herbaceous plants, ground covers, and root crops. (2) **Water Harvesting: Swales** on contour to direct water to planting areas. **Ponds** for irrigation and aquaculture. (3) **Natural Pest Control**: Companion planting (e.g., marigolds to deter nematodes). Plant diversity to break pest cycles. (4) **Windbreaks**: Rows of fast-growing trees like bamboo or moringa on farm boundaries. (5) **Seasonal Crop Rotation**: Alternating legumes, cereals, and cover crops for balanced nutrient use.

Methods of Farm Landscape Design:

(1) **Selection of Compatible Plant and Animal Species: Complementary Roles**: Choose plant species that provide shelter, food, or habitat for animals, while also selecting animals that can help with pest control, soil aeration, or nutrient cycling. For example: Chickens can be integrated with vegetable crops to control pests

and weeds. Cattle can graze on grasses while providing manure that can be used to fertilize the soil. Bees are essential for pollination, benefiting both fruit trees and vegetables. **Diverse Planting:** Consider planting a variety of crops (fruit trees, vegetables, legumes, cover crops) that can support different animal species. This diversity reduces the risk of pest infestations and increases the farm's ecological resilience. (2) **Spatial Arrangement and Design: Zoning:** Design the farm to create zones for plants and animals that allow for efficient use of space and resources. For example, livestock may be kept in areas where grazing doesn't interfere with the growth of high-value crops, while also benefiting from plant residues or protection from harsh weather. **Agroforestry:** Planting trees alongside crops or grazing areas can offer shade for animals, create microclimates, and provide additional sources of income through timber or fruit. Trees also act as windbreaks, protecting crops and animals from extreme weather.

6.4.6 Designing Multilayer Polycropping System:

Each individual farmer should design their own unique multilayer polycropping system for natural farming, according to their own need and circumstances.

Objectives of Multilayer Polycropping System:

Here are in brief the objectives of Multilayer polycropping systems: (1) **Cropping Intensity:** To maximise the cropping intensity of the natural farm by accommodating as many types of crops as possible. (2) **Solar Energy Harvesting:** To maximise the leaf area index (LAI) of the agroecosystem, to maximise the harvesting of solar energy. (3) **Atmospheric Water Harvesting:** To maximise the capacity of the agroecosystem to harvest atmospheric water such as dew and fog, to maximise self-irrigation capacity. (4) **Rainwater Harvesting:** To maximise the rainwater harvesting capacity of the natural farming agroecosystem. (5) **Wind Break**: To protect the crops from the harm caused by high wind velocity. (6) **Biological Nitrogen Fixation**: To maximise the biological nitrogen fixation capacity of the natural farming agroecosystem through symbiosis (Rhizobium) and mutualism (PGPR). (7) **Phosphate Solubilisation**: To maximise phosphate solubilisation in soil through mutualism (AMF and PGPR). (8) **Natural Control**: To maximise natural control of insect pests, diseases and weeds of crops through interspecific interactions: (9) **Biomass Synthesis**: To maximise biomass synthesis and carbon sequestration in the natural farming agroecosystem. (10) **Aggregate Yield**: To maximise aggregate yield of crops. (11) Eco-Agritourism: To Maximise the potential for eco-agritourism traffic in the natural farming agroecosystem. (12) **Profit Maximisation**: The ultimate objective of designing the multilayer polycropping system is to maximise the profit of the Farmers.

Method of Designing Multilayer Polycropping System:

Designing a multilayer polycropping system for natural farming involves careful consideration of various factors to optimize resource use, increase biodiversity, and promote sustainability. This system mimics the structure of natural ecosystems by layering crops of different heights, light requirements, and root systems. (1) **Climate and Weather: Sunlight**: Choose crops based on their light requirements. Tall plants (e.g., trees) should allow enough light for the understory. **Rainfall**: Select drought-resistant or water-intensive crops depending on rainfall patterns. **Temperature**: Ensure that crops are suitable for local temperature ranges. (2) **Soil Characteristics: Fertility**: Match crops to the soil's nutrient profile. Use nitrogen-fixing plants like legumes to enhance fertility. **Texture and Drainage**: Ensure good soil drainage to avoid waterlogging in deeper layers. **pH**: Choose crops adapted to the soil's pH level or amend the soil with natural inputs like lime or organic compost.

(3) **Layering Design: Canopy Layer:** Tall trees (e.g., mango, coconut) provide shade and wind protection. **Sub-Canopy Layer:** Medium-height trees (e.g., guava, moringa) that thrive under partial sunlight. **Shrub Layer:** Bushy crops (e.g., coffee, pomegranate) that require less sunlight. **Herbaceous Layer:** Vegetables, herbs, and medicinal plants (e.g., turmeric, basil). **Ground Cover:** Low-growing plants (e.g., sweet potatoes, strawberries) to suppress weeds and retain moisture. **Root Layer:** Root crops (e.g., ginger, carrots) that grow underground, utilizing deeper soil layers. (4) **Crop Compatibility: Complementary Growth:** Choose crops that benefit each other, such as nitrogen-fixing legumes and heavy feeders like cereals. **Allelopathy:** Avoid plants that release chemicals inhibiting the growth of other crops (e.g., black walnut). **Pest and Disease Resistance:** Mix crops to reduce vulnerability to pests and diseases. (5) **Resource Utilization: Water:** Use drip irrigation or rainwater harvesting to efficiently distribute water across layers. **Nutrients:** Incorporate organic mulches and compost to nourish the soil and minimize competition for nutrients. **Light:** Ensure that all layers receive adequate sunlight based on crop requirements. (6) **Biodiversity and Ecosystem Services:** Include flowering plants to attract pollinators and pest predators. Introduce nitrogen-fixing species (e.g., beans, peas) to enhance soil health. Integrate multi-purpose plants (e.g., moringa for food, fodder, and soil improvement). (7) **Seasonal Planning:** Select crops with staggered planting and harvesting times to ensure continuous production. Rotate crops seasonally to maintain soil health and reduce pest buildup. (8) **Space Optimization:** Use vertical space efficiently by stacking crops with different growth habits. Adjust spacing to prevent overcrowding and competition. (9) **Water Management:** Design swales, trenches, or ponds to capture and store rainwater. Plant water-loving crops near water sources and drought-resistant crops in upland areas. (10) **Organic Input Availability:** Use locally available organic materials like cow dung, green manure, and crop residues. Avoid chemical

fertilizers and pesticides; rely on natural pest repellents like neem oil. (11) **Socio-Economic Factors: Market Demand:** Grow crops with high local demand and market value. **Labor Requirements:** Choose crops that match available labor capacity for maintenance and harvest. **Cost:** Ensure cost-effectiveness by utilizing existing farm resources. (12) **Long-Term Sustainability: Erosion Control:** Use ground covers and contour planting to prevent soil erosion. **Carbon Sequestration:** Include trees and perennials that store carbon and enhance resilience. **Regeneration:** Plan crop rotations and fallow periods for soil recovery.

CHAPTER 7

WATER CYCLE MANAGEMENT

7.1 Water Cycle in Natural Farming Agroecosystem:

The water cycle in an agroecosystem is a dynamic process influenced by agricultural practices, landscape features, climate, and soil characteristics. It involves the movement, storage, and transformation of water through various pathways, including precipitation, infiltration, evaporation, transpiration, and runoff. Below is a detailed characterization of the water cycle in an agroecosystem:

Components of Water Cycle in Natural Farming Agroecosystem:

(a) **Precipitation**: Source: Rainfall, snowfall. It is the primary input of water to the system. Influences soil moisture, plant growth, and groundwater recharge.

(b) **Infiltration**: Process: Water penetrates the soil surface and moves downward into the soil profile. Factors affecting infiltration: Soil texture and structure. Presence of vegetation or crop cover. Agricultural practices like plowing or mulching.

(c) **Runoff: Surface runoff**: Water flows over the soil surface to nearby streams or ponds. Enhanced by factors like: Compact soil., Sloped terrain. Lack of vegetation or crop cover.

(d) **Soil Moisture Storage**: Water retained in the soil is available for plant uptake. Influenced by: Soil type (clay, loam, or sandy soils). Organic matter content. Depth and drainage capacity of soil.

(e) **Evaporation**: Loss of water from soil and water bodies to the atmosphere as vapor. Influenced by temperature, humidity, wind speed, and soil cover.

(f) **Transpiration**: Loss of water from plants through their stomata during photosynthesis. Together with evaporation, it forms evapotranspiration, a key process in agroecosystems.

(g) **Percolation and Groundwater Recharge**: Excess water infiltrates deeper soil layers, reaching aquifers. Contributes to long-term water availability for irrigation.

(h) **Irrigation**: Artificial water input to compensate for insufficient precipitation.

(i) **Crop Water Uptake**: Plants absorb water through their roots for physiological processes, including photosynthesis and growth.

(j) **Drainage**: Excess water exits the root zone, either naturally or through artificial drainage systems.

(k) **Water Balance**: Balancing inputs (precipitation, irrigation) and outputs (evapotranspiration, runoff) is crucial for sustainable farming.

7.2 Objectives of Water Cycle Management in Natural Farming:

The objective of management of water cycle in natural farming agroecosystem, is to maintain optimum soil moisture condition (field capacity), automatically and permanently, to facilitate optimum plant growth and reproduction, without requiring any artificial irrigation to the crops (rainfed crops). The detailed objectives include:

7.2.1 Atmospheric Water Harvesting:

Some plants have the capacity to harvest atmospheric water such as fog and dew in their leaves and to transport the accumulated water from leaf surface to their root zone soil through specific channels in the leaves (Zulfiqar Ali et al. 2022). This is known as self-irrigation. For example, sugarcane and wheat. Some plants can do foliar absorption of dew.

7.2.2 Cloud Seeding and Inducing Rainfall:

Some plant species can emit biogenic volatile organic compounds (BVOC) or ice nucleating bacteria in the atmosphere during air turbulence that causes cloud seeding in the atmosphere and inducing rainfall. A combination of such rain trees can create a rainforest in the natural farming agroecosystem.

7.2.3 Rainwater Harvesting:

Plant canopies can absorb the shock of the high velocity of rain drops and can harvest rainwater with their leaves. Plant leaves, barks and dry leaves on the ground can absorb a good quantity of rainwater. Foliar absorption of rainwater is also very common.

7.2.4 Check Runoff and Rainwater Infiltration in Soil:

Dry leaves on the ground and mulching layer on the soil surface can check runoff of excess rainwater over the soil surface towards the direction of slope and can increase the rate of rainwater infiltration in soil. Earthworms and other burrowing insects and animals also help infiltration of rainwater in soil.

7.2.5 Water Holding Capacity of Soil:

Natural farming practices such as no till, no off-farm input, no irrigation, mulching, cover cropping, multilayer polycropping and mixed animal husbandry increases organic matter content in soil and increases microbial activity in biofilm formation and improvement of soil aggregates and soil structure. These factors increase water holding capacity in soil.

7.2.6 Evapotranspiration from Soil:

The plant biodiversity in the multilayer polycropping system creates a humid microclimate and the mulching layer in the soil surface effectively reduces the rate of evapotranspiration from the soil surface.

7.2.7 Reducing Water Requirement of Crops:

The water requirement in crops under natural farming agroecosystem can be vastly reduced by breeding drought resistant crops, and passively maintaining optimum soil moisture throughout the year under the Multilayer polycropping system.

7.2.8 Water Conservation:

Natural farming agroecosystems can maximise the efficiency of water conservation through biotic means.

7.2.9 Eliminating the Cost of Artificial Irrigation:

Natural farming agroecosystems can maximise water conservation naturally and help avoid artificial irrigation of crops that eliminate the cost of artificial irrigation.

7.3 Methods of Water Cycle Management in Natural Farming Agroecosystem:

Traditional farmers use various traditional methods in their own natural farming agroecosystems for efficient management of the natural water cycle. These techniques are unique in character and are completely different from the systems of irrigation used in modern agriculture (industrial monoculture). Here are a few common methods of water cycle management:

7.3.1 No Till:

The practice of no-till plays a pivotal role in managing water cycle in the natural farming agroecosystems, which include: (1) Soil Horizon: No-till restores natural soil horizons which is the natural filtration system for available water resources. (2) Soil Structure: No-till restores natural soil aggregates that are formed by microbial biofilms. These soil aggregates can store water and provide slow release of water to the plants. (3) Soil Organic Matter: No-till increases organic matter content in soil, which increases water holding capacity of the soil. (4) Soil Porosity: No-till restores earthworms, insects and other organisms in soil, which burrow in the soil and increases the soil porosity for improving the capacity of rainwater infiltration in soil. (5) Water Harvesting: No-till accumulates crop residues, weeds and other organic materials on the soil surface, which help rainwater harvesting. (6) Runoff: Crop residues and weeds check rainwater runoff. (7) Rainwater Infiltration: Crop residues and organic substrate on soil surface help infiltration of rainwater in soil. (8) Evapotranspiration: Crop residues, and other organic matter on soil surface act as natural mulching and reduces the rate of evapotranspiration from soil surface.

7.3.2 No Off-Farm Input:

Chemical fertilisers and herbicides cause chemical reactions in soil and reduce the water holding capacity of the soil. Complete avoidance of off-farm commercial inputs help water retention in soil.

7.3.3 No Irrigation:

Artificial Irrigation by using underground water often increases salinity, alkalinity or acidity in soil and reduces the capacity of rainwater infiltration in soil and reduces the water holding capacity of the soil. Natural farming agroecosystems automatically and efficiently manage the water content in soil at optimum level (field capacity) throughout the year for optimum plant growth. There remains no need for artificial irrigation in the natural farming agroecosystems.

7.3.4 Blue Infrastructure:

Farmers have traditionally developed many unique methods and techniques for construction of permanent water reservoirs and water transportation channels using locally available materials, for efficient harvesting and conservation of rainwater and for it's efficient use in agriculture.

7.3.5 Mulching:

Mulching is a traditional method of keeping the soil surface always covered with a thick layer of natural organic materials such as dry leaves or crop residues. Mulching is an important tool for water cycle management in the natural farming agroecosystem. (1) Control Splash: Mulching layer intercept the high velocity of the raindrops from the atmosphere and protect the soil surface from splash, displacement of soil particles, clogging of soil pores and reduction of the rate of rainwater infiltration in soil. (2) Control Runoff: Mulching layer on soil prevents runoff of rainwater towards the slope. (3) Improve Infiltration: Mulching layer on soil surface improves the rate of rainwater infiltration in soil. (4) Water Holding Capacity: The dry organic materials used in mulching can absorb an enormous amount (say 10 times it's dry weight) of water for a long period of time. (5) Evapotranspiration: Mulching layer on soil reduces the rate of evapotranspiration from soil surface which effectively reduces water loss from soil. (6) Field Capacity: Mulching layer on soil surface helps maintain optimum soil moisture level (field capacity) automatically, throughout the year to promote optimum plant growth and development without needing any irrigation water.

7.3.6 Cover Cropping:

Cover cropping is a traditional practice of growing crops all round the year, to keep the soil surface always (permanently) covered with green leaves. Cover cropping is an important tool for water cycle management in the natural farming agroecosystem. (1) Control Splash: Cover crop canopies intercept the high velocity of the raindrops from the atmosphere and protect the soil surface from splash, displacement of soil particles, clogging of soil pores and reduction of the rate of rainwater infiltration in soil. (2) Control Runoff: Cover crops prevents runoff of rainwater towards the slope of land. (3) Improve Infiltration: Cover crops improve the rate of rainwater infiltration in soil. (4) Water Holding Capacity: The organic crop residues form a mulching layer on the soil surface which can absorb rainwater for a long period of time. (5) Evapotranspiration: Cover crops reduce the rate of evapotranspiration from the soil surface which effectively reduces water loss from soil. (6) Field Capacity: Cover crops helps maintain optimum soil moisture level (field capacity) automatically, throughout the year, to promote optimum plant growth and development without needing any irrigation water.

7.3.7 Multilayer Polycropping:

Multilayer polycropping is the most fascinating component of management of the water cycle in the natural farming agroecosystem. (1) Atmospheric Water Harvesting (Self-Irrigation): Plants leaves can collect atmospheric water in the form of dew and fog. Some plant leaves can carry this water to their root zone soil, through their leaf channels. This is known as self irrigation. Example wheat and sugarcane. (2) Cloud Seeding: Some plants can emit biogenic volatile organic compounds (BVOC) and ice nucleating bacteria such as *Pseudomonas syringae*, from their leaf surface during air turbulence. These ingredients along with transpiration from plants can induce cloud seeding in the atmosphere. (3) Inducing Rainfall: Plant leaves can induce rainfall from clouds in the atmosphere. (4) Water Harvesting: The compact canopy of the multilayer polycropping system can effectively intercept the raindrops and help harvesting of rainwater in the agroecosystem. (5) Water Absorption: Plant leaves and barks can directly absorb significant quantities of rainwater. (6) Stop Runoff: Leaf litters under the plants actively resist rainwater runoff over the soil surface towards the slope of the land. (7) Increased Infiltration: Leaf litters form a mulch layer over the soil surface which effectively increase rainwater infiltration in soil. (8) Microclimate: The multilayer polycropping structure form a closed humid microclimate beneath their canopies which help conservation of water. (9) Evapotranspiration: The humid microclimate within the canopy architecture and the natural mulching layer on the soil surface effectively reduce the rate of evapotranspiration from the soil surface that help water conservation. (10) Water Conservation: Multilayer polycropping system is the best nature based solution (NbS) for rainwater harvesting and water conservation in the natural farming agroecosystem.

CHAPTER 8

SOIL HEALTH MANAGEMENT

8.1 Definitions:

It is important to define soil properties, soil fertility and soil health, to understand the role of soil health in natural farming agroecosystem.

8.1.1 Soil Properties:

Soil properties are critical characteristics that determine the soil's ability to support plant growth, water retention, and nutrient availability. These properties are classified into physical, chemical, and biological categories. Below is a detailed explanation of each type:

1. Physical Properties:

a. **Texture:** Refers to the relative proportions of sand, silt, and clay particles. **Sandy soils:** Coarse particles, low water retention. **Clay soils:** Fine particles, high water retention but poor drainage. **Loamy soils:** Balanced texture, ideal for agriculture.

b. **Structure:** Describes how soil particles aggregate into clumps (peds). **Granular structure:** Found in topsoil, promotes aeration and water infiltration. **Blocky or platy structure:** Found in subsoil, can restrict root growth and water movement.

c. **Porosity:** Refers to the volume of pores (spaces) within the soil. Affects water retention, infiltration, and air exchange. Soils with higher porosity (e.g., sandy soils) allow faster drainage.

d. **Bulk Density**: The mass of soil per unit volume, including pore spaces. **High bulk density**: Compacted soil, poor root growth. **Low bulk density**: Well-aerated, suitable for plant growth.

e. **Color**: Indicates organic matter content, drainage conditions, and mineral composition. **Dark soils**: Rich in organic matter. **Red or yellow soils**: High iron content, well-drained. **Grey soils**: Poorly drained or waterlogged.

f. **Water-Holding Capacity**: The soil's ability to retain water for plant use. **Sandy soils**: Low capacity. **Clay soils**: High capacity but less available water for plants.

g. **Permeability**: The rate at which water and air move through soil. **Sandy soils**: High permeability. **Clay soils**: Low permeability.

h. **Consistency**: Describes soil's resistance to deformation and rupture. Dry, moist, and wet conditions affect consistency.

2. Chemical Properties:

a. **pH**: Indicates soil acidity or alkalinity on a scale of 1–14. **Acidic soils**: pH < 7, common in high rainfall areas. **Neutral soils**: pH = 7, ideal for most crops. **Alkaline soils**: pH > 7, often rich in calcium and magnesium.

b. **Cation Exchange Capacity (CEC)**: Measures soil's ability to hold and exchange positively charged ions (cations) like calcium (Ca^{2+}), magnesium (Mg^{2+}), and potassium (K^+). **High CEC**: Fertile soils with good nutrient-holding capacity. **Low CEC**: Sandy soils with low fertility.

c. **Nutrient Content: Macronutrients:** Nitrogen (N), Phosphorus (P), Potassium (K).

Micronutrients: Iron (Fe), Zinc (Zn), Copper (Cu), etc. Nutrient availability depends on soil type, organic matter, and pH.

d. **Salinity**: Refers to the concentration of soluble salts in soil. **High salinity**: Hinders plant growth due to osmotic stress.

e. **Organic Matter**: Decomposed plant and animal material that improves soil fertility, water retention, and microbial activity.

f. **Base Saturation**: Proportion of soil bases (Ca, Mg, K, Na) relative to total CEC. Higher base saturation indicates fertile soil.

3. Biological Properties:

a. **Microbial Activity**: Soil is home to bacteria, fungi, protozoa, and algae that decompose organic matter and cycle nutrients. **Beneficial organisms**: Rhizobia (nitrogen fixation), Mycorrhizae (enhance nutrient uptake).

b. **Soil Organic Carbon**: Indicator of organic matter, affects soil structure, water retention, and nutrient cycling.

c. **Soil Fauna**: Earthworms, insects, and burrowing animals improve aeration and nutrient distribution.

d. **Enzymatic Activity**: Soil enzymes, like urease and phosphatase, are crucial for nutrient transformations.

e. **Biological Diversity**: High diversity promotes a resilient soil ecosystem and sustainable productivity.

4. Integrated Properties

a. **Soil Fertility**: The ability of soil to provide essential nutrients to plants in sufficient quantities.

b. **Soil Erosion:** The susceptibility of soil to be eroded by wind or water, influenced by texture, structure, and vegetation cover.

c. **Soil Compaction:** Reduced porosity and increased bulk density due to heavy machinery or livestock, restricting root growth and water movement.

d. **Soil Resilience:** The soil's ability to recover from disturbances like drought, flooding, or intensive agriculture.

8.1.2 Soil Ecology:

Soil Ecology is the study of the interactions between soil organisms, their environment, and the processes that drive nutrient cycling, organic matter decomposition, and energy flow within the soil ecosystem. It is a vital aspect of understanding soil health, fertility, and its role in supporting plant life and broader ecosystems. Key Components of Soil Ecology

Soil Organisms:

Soil harbors a diverse range of organisms that interact with each other and their environment. These organisms are classified into three groups based on size: (1) **Microorganisms** (less than 0.2 mm): **Bacteria:** Decompose organic matter, fix nitrogen (e.g., Rhizobium), and cycle nutrients. **Fungi:** Break down complex organic compounds; mycorrhizal fungi enhance plant nutrient uptake. **Algae:** Perform photosynthesis, adding organic carbon to the soil. **Protozoa:** Regulate bacterial populations and release nutrients. (2) **Mesofauna** (0.2–2 mm): Includes mites, springtails, and nematodes. Decompose organic matter and contribute to soil structure. (3) **Macrofauna** (greater than 2 mm): **Earthworms:** Aerate soil, mix organic and mineral components, and enhance nutrient availability. **Ants and termites:** Improve soil structure and redistribute organic matter.

Soil Organic Matter (SOM):

Organic material derived from plant and animal residues. Plays a key role in nutrient cycling, soil structure, and water retention. Acts as an energy source for soil organisms.

Soil Food Web:

Represents the feeding relationships among soil organisms. Primary producers (plants and algae) provide energy, which is transferred to decomposers (bacteria and fungi), and then to higher trophic levels (predators like nematodes and mites).

Nutrient Cycling:

Soil organisms drive the cycling of essential nutrients such as carbon (C), nitrogen (N), phosphorus (P), and sulfur (S). **Nitrogen Cycle:** Includes nitrogen fixation, ammonification, nitrification, and denitrification. **Carbon Cycle:** Organic carbon decomposition releases CO_2 and contributes to SOM formation. **Phosphorus and Sulfur Cycles:** Microorganisms solubilize phosphorus and transform sulfur compounds.

Soil Structure and Aggregation:

Organisms like fungi and earthworms contribute to the formation of soil aggregates. Aggregates enhance porosity, water infiltration, and resistance to erosion.

Processes in Soil Ecology:

Decomposition: Breakdown of organic matter into simpler compounds by soil organisms. Releases nutrients in forms available to plants (e.g., ammonium, nitrate, phosphate). **Bioturbation:** Mixing of soil layers by macrofauna like earthworms and termites. Improves aeration, water infiltration, and organic matter distribution. **Symbiosis: Mycorrhizae:** Fungal associations with plant roots improve water and nutrient uptake. **Nitrogen–Fixing Bacteria:** Convert atmospheric nitrogen into forms usable by plants (e.g., Rhizobium in legume root nodules). **Pathogenesis and Pest Control:** Some soil organisms cause plant diseases (e.g., pathogenic fungi), while others suppress pathogens and pests (e.g., predatory nematodes).

8.1.3 Soil Health:

There is no standard definition of soil health in soil science. The term "health" refers to the condition or state of a living organism. Natural soil is composed of both non-living and living components. Due to the presence of living organisms, natural soil, as a whole, behaves like a living organism. Therefore soil health can be defined as the health conditions of the living organisms in soil, such as plants, animals, and microorganisms. Soil health varies with time and environmental conditions. Soil health cannot be tested in the laboratory by analysing dead soil samples. Soil health can only be tested in the field by observing the health conditions of indicator plants (eg. sunflower), indicator animals (earthworms) and indicator microorganisms (wild mushrooms) that occur in a particular location, and at a particular time. Natural farming is concerned only about soil health, because soil health can modify all other soil properties for crop growth.

8.2 Objectives of Soil Health Management:

Objectives of soil health management in natural farming (polyculture) are completely different from the objectives of soil fertility management in conventional agriculture (monoculture). Natural farming aims to maintain a high level of soil health, automatically and permanently, by providing an ideal habitat for soil organisms. Whereas, conventional agriculture intends to destroy soil ecosystem, soil biodiversity, ecosystem services and soil health, to maximise the use of synthetic chemical fertilisers, micronutrients and plant growth regulators. Here are main objectives of soil health management in natural farming:

8.2.1 Restoration and Conservation of Soil Horizons:

Soil horizons are distinct layers within the natural soil profile, each with unique characteristics. The primary horizons include: **O Horizon:** The topmost layer, rich in organic matter like decomposed leaves and humus, often dark in color. O horizon is the most important layer of soil for natural farming because this layer provides the food and habitat for the biodiversity of soil organisms. **A Horizon (Topsoil):** Located beneath the O horizon, it contains minerals and organic material, supporting most plant life and biological activity. **E Horizon:** A leached layer that shows significant loss of clay and nutrients, often lighter in color; common in forested areas. **B Horizon (Subsoil):** Accumulates materials from upper layers, denser and less fertile than topsoil. **C Horizon:** Composed of weathered rock or parent material, it is less affected by soil-forming processes. **R Horizon:** The bedrock layer beneath the soil profile.

8.2.2 Restoration and Conservation of Soil Structure:

Soil structure refers to the arrangement of soil particles into aggregates, known as peds. These aggregates vary in size and shape, influencing water movement, air circulation, and root growth. Key components include: **Aggregates:** Groups of soil particles bound together, affecting stability and porosity. Well-structured soils have a mix of small and large aggregates. **Pores:** Spaces between aggregates that allow for air and water flow; categorized as micro, meso, and macro pores. **Formation Processes:** Soil structure develops through physical (e.g., freezing/thawing), chemical (ionic bonding), and biological (organic matter and microbial activity) processes.

8.2.3 Restoration and Conservation of Soil Organic Matter:

Soil organic matter (SOM) consists of decomposed plant and animal materials, including living organisms and their byproducts. It plays a crucial role in soil fertility and health by providing essential nutrients like nitrogen, phosphorus, and sulfur, as well as micronutrients. SOM is categorized into three main pools: fresh plant residues, active organic matter, and stable organic matter (humus) that contributes to long-term nutrient storage and soil structure improvement. Additionally, it enhances water retention, aeration, and microbial activity, making it vital for sustainable agricultural practices.

8.2.4 Restoration of Water Holding Capacity of Soil:

Soil water holding capacity (SWHC) refers to the amount of water soil can retain for plant use, influenced primarily by soil texture and organic matter. Key points include: **Soil Texture:** Soils with smaller particles, like clay and silt, have higher water holding capacities compared to sandy soils. For example, silt loam can hold about 2.4 inches of water per foot, while sand holds only 0.8 inches. **Organic Matter:** Increases SWHC by enhancing soil structure and moisture retention. Higher organic matter content allows soils to hold more water. **Field Capacity:** Represents the maximum amount of water retained after excess water has drained; maintaining this level is crucial for optimal crop production.

8.2.5 Controlling Soil Erosion, Soil Degradation and Desertification of Land:

Soil erosion is the process of removing the upper layer of soil, primarily caused by natural forces like solar energy, water and wind, as well as human activities. Key causes include: **Rainfall and Flooding:** Intense rainfall leads to various types of erosion, including splash, sheet, rill, and gully erosion, which wash away topsoil into water bodies. **Agricultural Practices:** Poor farming techniques, such as ploughing along slopes, disrupt soil structure and increase vulnerability to erosion. **Vegetation Loss**: Removal of plants exposes soil to erosive forces; grazing animals further exacerbate this by uprooting vegetation. **Wind**: Particularly in arid regions, strong winds can lift and transport soil particles, leading to significant erosion.

8.2.6 Restoration of Soil Biodiversity:

Soil biodiversity refers to the variety of life forms present in the soil, including plants (flora), microorganisms (bacteria, fungi), mesofauna (springtails, mites), and macrofauna (earthworms, insects). Soil biodiversity is crucial for: **Nutrient Cycling**: Essential for the breakdown of organic matter and nutrient availability for plants. **Ecosystem Services**: Supports water purification, carbon sequestration, and pest suppression. **Plant Growth**: Enhances soil structure and improves water retention, benefiting root development.

8.2.7 Restoration of Plant Health and Immunity:

Plant health is influenced by several critical factors that ensure optimal growth and development. Key factors include: **Water**: Essential for nutrient uptake and photosynthesis; both overwatering and drought can harm plants. **Nutrients**: Plants require 17 essential nutrients, with deficiencies leading to stunted growth and poor health. Nutrients are primarily absorbed through water. **Light**: Adequate light is necessary for photosynthesis; insufficient light can hinder growth, especially root development. **Temperature**: Affects metabolic processes; extreme temperatures can slow growth or cause stress. **Relative Humidity**: Ideal humidity levels support metabolic processes, while extremes can lead to water loss or fungal diseases.

8.2.8 Improving Carbon Sequestration in Soil:

Soil carbon sequestration (SOC) is the process of capturing atmospheric CO_2 and storing it in the soil as organic carbon. This occurs primarily through photosynthesis, where plants convert CO_2 into organic compounds, which then enter the soil via plant residues and root exudates.

8.2.9 Bioremediation of Soil Pollution:

Bioremediation of soil pollution is a sustainable method that utilizes living organisms, primarily microorganisms and plants, to degrade, transform, or detoxify contaminants in the soil. This approach is environmentally friendly and cost-effective compared to conventional remediation methods. Key techniques include: **Phytoremediation**: Involves using plants to absorb and detoxify pollutants. **Bioaugmentation**: Introduces specific microorganisms to enhance degradation of contaminants. **Biostimulation**: Adds nutrients to stimulate indigenous microbial populations for enhanced pollutant breakdown.

8.2.10. Minimising Risk and Cost of Production of Crops:

Natural farming improves soil health automatically which reduces the risk of crop loss due to environmental factors and due to insect pests, diseases and weeds. Natural farming avoids disturbance of the soil ecosystem which reduces the cost of production of crops.

8.2.11. Maximising Yield, Quality, Price and Profit of the Farmers:

The ultimate objective of soil health management in natural farming is to minimise the risk and cost of production of crops and to maximise the aggregate yield, product quality, product price and to maximise the profit of the farmers.

8.3 Methods of Soil Health Management:

Farmers use traditional agricultural knowledge and traditional practices for management of soil health in the natural farming agroecosystems. These traditional practices essentially restore and maintain the natural soil ecosystem and restore the habitat for soil biodiversity. Here are the common methods of soil health management.

8.3.1 No Till:

No-till builds the foundation of the natural farming agroecosystems. No-till farming is a sustainable agricultural practice that offers numerous advantages, particularly in terms of soil health, environmental sustainability and economic efficiency. Here are the key benefits:(1) **No-till stops soil erosion**: Avoiding tillage and soil disturbance accumulate crop residues on the soil surface. Crop residues protect the soil surface from soil erosion caused by solar heat, rainwater and high wind velocity. (2) **No-till resume the process of soil formation (pedogenesis)**: Soil formation is the process through which soil develops from parent material, influenced by various factors including climate, organisms, topography, and time. Key processes include: **Weathering**: Physical (e.g., temperature changes), chemical (e.g., reactions with water), and biological (e.g., plant roots) breakdown of rocks. **Climate**: Temperature and moisture affect weathering rates and organic matter decomposition, with warmer, wetter climates promoting faster soil development. **Biological Activity**: Organisms contribute to soil formation by breaking down materials and enhancing nutrient cycling

(3) **No-till restores soil horizons**: Soil horizons are distinct layers within a natural soil profile, each with unique properties and compositions. The main horizons include: **O Horizon**: Composed of organic matter like decomposing leaves, varying in thickness. **A Horizon (Topsoil)**: Rich in minerals and organic material, crucial

for plant growth. **E Horizon:** Characterized by leaching, it often appears bleached due to the loss of clay and nutrients. **B Horizon (Subsoil):** Accumulates minerals from above layers, typically denser and less fertile. **C Horizon:** Consists of weathered rock and parent material, less affected by soil formation processes. **R Horizon:** Bedrock layer beneath the soil profile. Among these soil horizons, O-Horizon is the most important for natural farming, because it provides food and habitat to the soil biodiversity. Tillage completely destroys these soil horizons and converts it into an amorphous mass of dust (dust bowl). No-till can restore and conserve these horizons over time. (4) **No-till restores soil structure (soil aggregates):** No-till avoids disruption of soil aggregates, promoting better aeration and water infiltration. (5) **No-till increases organic matter content in soil (carbon sequestration):** Soil carbon sequestration (SOC) is the process of capturing atmospheric CO_2 and storing it in the soil as organic carbon, primarily through photosynthesis by plants. This carbon enhances soil health by improving structure, nutrient retention, and water capacity, which can increase agricultural productivity. (6) **No-till improves water holding capacity of soil:** Avoiding soil disturbance increases crop residues and organic matter content on the soil surface.. Crop residues act as mulch, increasing rainwater harvesting, increasing rainwater infiltration in soil, reducing evaporation and improving soil moisture levels. (7) **No-till reduces runoff:** Avoiding tillage increases organic matter content on the soil surface which enhances infiltration of rainwater in soil and reduces water loss (runoff) during heavy rains. (8) **No-till restores soil biodiversity:** Avoiding soil disturbance increases organic matter content in soil and increases soil moisture level. In this way no-till provides food and habitat for soil biodiversity and supports a thriving ecosystem of soil microorganisms, earthworms, and beneficial insects. (9) **No-till restores soil ecology:** Soil ecology is the study of interactions among soil organisms and their environment, focusing on nutrient cycling, soil structure, and biodiversity. (10) **No-till restores ecosystem services:** Soil provides essential ecosystem services that support human well-being and environmental health. Key services include: **Provisioning Services:** Production of food, fiber, and fuel. **Regulating Services:** Water filtration, climate regulation, and erosion control. **Cultural Services:** Recreational and aesthetic benefits, contributing to cultural identity. **Supporting Services:** Nutrient cycling, soil formation, and habitat provision for biodiversity. (11) **No-till reduces greenhouse gas emission:** Eliminating mechanised tillage decreases fossil fuel usage, reducing carbon emissions. (12) **No-till escapes soil compaction:** Elimination of mechanised tillage avoids the chance of soil compaction and formation of hard impervious crust in the subsurface soil.

8.3.2 No Off-Farm Input:

Farmers should avoid all sorts of commercial agricultural inputs in their natural farming agroecosystem to rebuild soil health and to save huge amounts of money as input costs. Common inputs used in conventional agriculture includes:

Hybrid Seed: Avoiding hybrid seeds and opting for traditional or open-pollinated seeds offers several benefits, particularly in terms of sustainability, biodiversity, and long-term farming resilience. Here are the key advantages: (1) **Seed Sovereignty: Self-Reliance:** Farmers can save and reuse seeds from open-pollinated varieties, reducing dependence on seed companies. **Cost Savings:** Avoids the need to purchase new seeds every season, as hybrid seeds do not produce true-to-type offspring. (2) **Biodiversity Conservation: Genetic Diversity:** Traditional seeds maintain a wider range of genetic traits, making crops more adaptable to changing environmental conditions. **Local Adaptation:** Indigenous seeds are often better suited to local climates and soil conditions, requiring fewer inputs like fertilizers and pesticides. (3) **Sustainability:**

Resilience to Pests and Diseases: Traditional seeds often have natural resistance, developed over generations, to local pests and diseases. (4) **Nutrition and Flavor: Better Nutritional Value**: Many traditional varieties are richer in nutrients compared to hybrids, which are often bred for yield over quality. **Superior Taste**: Traditional crops are often prized for their authentic and diverse flavors, which may be diminished in hybrids bred for uniformity. (5) **Environmental Benefits: Adaptability to Natural Farming**: Traditional seeds often perform better in organic systems, which emphasize natural processes and minimal chemical use. **Pollinator-Friendly**: Open-pollinated crops support pollinator populations better, as they are not engineered for uniform flowering patterns. (6) **Economic and Cultural Preservation: Support for Small-Scale Farmers**: Promotes farming practices that are less reliant on expensive inputs, benefiting small and marginal farmers. **Preservation of Heritage**: Maintains agricultural traditions and indigenous knowledge, ensuring cultural continuity. (7) **Resilience to Climate Change: Natural Adaptability**: Open-pollinated seeds have a broader genetic base, making them more adaptable to climate variability. **Local Selection**: Farmers can select and save seeds that thrive in specific conditions, enhancing long-term resilience.

Chemical Fertilisers: Avoiding chemical fertilizers and adopting natural farming practices has significant advantages for soil health, crop quality, the environment, and overall sustainability. Below are the key benefits: (1) **Improved Soil Health: Enhanced Microbial Activity**: Natural fertilizers like compost and manure support beneficial soil microorganisms that improve soil structure and fertility. **Maintains Soil Structure**: Avoids the compaction and degradation often caused by chemical salts in synthetic fertilizers. **Increases Organic Matter**: Promotes the build-up of humus, enhancing water retention and nutrient availability. (2) **Better Crop Quality: Higher Nutritional Value**: Crops grown without chemical fertilizers often have higher levels of vitamins, minerals, and antioxidants. **Natural Taste and Flavor**: Produce tends to be tastier and more authentic compared to chemically-fertilized crops. (3) **Environmental Benefits: Prevents Water Pollution**: Avoids runoff of synthetic nutrients into water bodies, reducing risks of algal blooms and eutrophication. **Reduces Greenhouse Gas Emissions**: Organic alternatives generally have a lower carbon footprint compared to the production and application of chemical fertilizers. **Preserves Biodiversity**: Encourages diverse ecosystems by minimizing chemical disruption to soil, water, and surrounding habitats. (4) **Economic Advantages: Cost Savings**: Reduces the expense of purchasing chemical fertilizers and mitigates long-term costs of soil rehabilitation caused by overuse of chemicals. **Local Resource Utilization**: Encourages the use of locally available organic materials like manure, compost, and green manure. (5) **Human and Animal Health: Reduces Exposure to Harmful Chemicals**: Minimizes the risk of health issues caused by chemical residues in food and water. **Safer Farming Environment**: Protects farmworkers and livestock from exposure to toxic substances. (6) **Resilience to Climate Change: Improves Water Retention**: Organic matter in the soil helps crops withstand droughts and heavy rains better. **Supports Carbon Sequestration**: Builds organic carbon levels in soil, helping combat climate change.

Synthetic Plant Growth Regulators: Avoiding synthetic plant growth regulators (PGRs) in agriculture can lead to healthier crops, more sustainable farming practices, and reduced environmental impact. Here are the key advantages: (1) **Healthier Crops and Consumers: No Chemical Residues**: Eliminates the risk of harmful PGR residues in food, ensuring safer produce for consumers. **Better Nutritional Value**: Naturally grown crops often retain more vitamins, minerals, and antioxidants. (2) **Environmental Benefits: Preserves Ecosystem Balance**: Avoids disrupting soil microbes, beneficial insects, and other organisms critical to a healthy ecosystem. **Reduces Soil and Water Contamination**: Prevents leaching of synthetic chemicals into groundwater and surface water. **Protects Pollinators**: Minimizes harm to bees and other pollinators, which can be affected by synthetic chemicals. (3) **Enhanced Soil Health: Maintains Natural Rhythms**: Allows plants to grow at their natural pace, which supports soil nutrient cycling and organic matter buildup. **Prevents Soil Degradation**: Avoids the side effects of synthetic chemicals, such as reduced soil fertility over time. (4) **Economic and Farming Advantages: Cost Savings**: Reduces reliance on expensive synthetic inputs, lowering farming costs. **Encourages Local Inputs**: Promotes the use of natural alternatives like compost teas, seaweed extracts, or microbial inoculants, which are often locally available. **Empowers Farmers**: Avoids dependency on agribusiness corporations for synthetic PGRs. (5) **Sustainable Farming Practices: Promotes Biodiversity**: Natural farming methods enhance on-farm biodiversity, supporting resilient agro-ecosystems. **Encourages Traditional Knowledge**: Reinforces the use of time-tested practices like crop rotation, intercropping, and the use of biofertilizers. (6) **Long-Term Crop Resilience: Natural Growth Processes**: Plants grown without synthetic PGRs develop stronger roots and better adapt to environmental stresses. **Resilience to Climate Change**: Encourages natural traits that help crops withstand drought, heat, or flooding.

Organic Fertilisers: Organic fertilisers and composts are commonly used in organic farming as an alternative input for plant growth. Natural farming, however, avoids all sorts of inputs, either chemical, organic or biological, to avoid cost and contamination. Natural farming totally relies on in-situ regeneration of fertilisers and pesticides by native soil microorganisms.

Biofertilisers: Biofertilisers are commercial cultures of a particular species of alien invasive species of microorganism, which can improve biological nitrogen fixation or phosphate solubilisation in soil. Avoiding use of biofertilizers in natural farming agroecosystems has many benefits. (1) **Environmental Sensitivity**: Biofertilizers are dependent on favorable environmental conditions, such as specific temperatures, moisture levels, and pH, for optimal performance. (2) **Ineffective in Poor Soils**: Their efficacy can be reduced in soils with low organic matter or unfavorable microbial environments. (3) **Short Shelf Life**: Many biofertilizers have a limited shelf life due to the living microorganisms they contain. (4) **Special Storage Requirements**: They often need refrigeration or cool storage to maintain viability, increasing logistical complexity. (5) **Specific Crop and Soil Requirements**: Certain biofertilizers are effective only with specific crops or soil types, limiting their versatility. (6) Risk of Contamination: **Introduction of Pathogens**: Improperly processed or contaminated biofertilizers can introduce harmful pathogens into the soil. (7) **Unintended Species**: There is a risk of introducing non-native or invasive microbial species that could disrupt the local soil ecosystem. (8) **Inconsistent Performance**: The effectiveness of biofertilizers can vary based on field conditions, crop types, and application methods. (9) **Competition with Native Microorganisms**: Introduced microorganisms might face competition from native soil microbes, reducing their efficacy. (10) **Higher Initial Costs**: While biofertilizers may reduce costs in the long term, their initial investment can be higher due to production, storage, and transportation requirements.

Vermicompost: While vermicompost is generally considered as beneficial for agriculture, it has many harmful effects on natural farming agroecosystems. Use of vermicompost increases the cost of cultivation significantly. Vermicomposts can introduce alien invasive species of earthworms through their cysts. Vermicompost may introduce other alien invasive microorganisms or pathogenic microorganisms in soil. Vermicompost may cause biomagnification of heavy metals and other substances.

Herbicides: Herbicides, while effective in controlling weeds and improving crop yields, have several disadvantages that can impact the environment, human health, and long-term agricultural sustainability. Below are the key drawbacks: (1) **Environmental Impact:** (a) **Soil Degradation:** Prolonged use of herbicides can harm beneficial soil organisms, reducing soil fertility and structure. (b) **Water Pollution:** Herbicides can leach into groundwater or run off into water bodies, contaminating aquatic ecosystems. (c) **Harm to Non-Target Species:** Herbicides can unintentionally harm beneficial plants, insects, and animals. (2) **Health Risks:** (a) **Toxicity to Humans:** Prolonged exposure to certain herbicides is associated with skin irritation, respiratory problems, hormonal disruptions, and even cancer. (b) **Residues in Food:** Improper use can result in herbicide residues on crops, posing health risks to consumers. (3) **Weed Resistance:** (a) **Development of Superweeds:** Over-reliance on herbicides can lead to the emergence of herbicide-resistant weeds, making them harder to control over time. (b) **Increased Costs:** Resistant weeds may require higher doses or additional herbicides, escalating costs for farmers. (4) **Biodiversity Loss:** (a) **Reduction in Plant Biodiversity:** Herbicides can eliminate native plants and wildflowers, reducing biodiversity in agricultural landscapes. (b) **Impact on Pollinators:** Herbicides can indirectly harm pollinators like bees and butterflies by reducing the availability of flowering plants. (5) **Economic Costs:** (a) **High Cost of Herbicides:** The continuous purchase of herbicides can be expensive, particularly for small-scale farmers. (b) **Increased Dependency:** Over time, reliance on herbicides can lead to reduced adoption of sustainable weed management practices. (6) **Soil and Crop Health:** (a) **Chemical Residue Build-Up:** Repeated use can lead to accumulation of harmful chemicals in the soil, affecting subsequent crops. (b) **Stunted Crop Growth**: Non-target damage from herbicides can result in reduced crop vigor and yield.

Pesticides: The use of chemical pesticides in agriculture, while effective in controlling pests and improving yields, has numerous disadvantages that can affect human health, the environment, and long-term agricultural sustainability. Below are the key drawbacks: (1) **Environmental Damage**: (a) **Soil Degradation**: Chemical pesticides can kill beneficial soil organisms, reducing soil fertility and disrupting the natural nutrient cycle. (b) **Water Pollution**: Pesticides can leach into groundwater or run off into rivers and lakes, harming aquatic ecosystems. (c) **Air Contamination**: Pesticide drift during application can pollute the air, affecting nearby ecosystems and communities. (2) **Harm to Non-Target Species**: (a) **Impact on Pollinators**: Pesticides can harm bees, butterflies, and other pollinators essential for crop production. (b) **Biodiversity Loss**: Non-target plants and animals, including beneficial insects and natural pest predators, may be adversely affected. (3) **Human Health Risks**: (a) **Toxicity to Farmers and Workers**: Prolonged exposure to pesticides can cause acute and chronic health issues, including respiratory problems, skin irritation, neurological disorders, and cancer. (b) **Residues in Food**: Improper application can lead to pesticide residues on food, posing risks to consumers. (4) **Development of Pest Resistance**: (a) **Pesticide Resistance**: Over time, pests can develop resistance to specific chemicals, making them less effective and requiring higher doses or newer pesticides. (b) **Emergence of Super-Pests**: Resistant pests can become more difficult and costly to manage. (5) **Economic Costs**: (a) **High Input Costs**: Chemical pesticides can be expensive, especially for small-scale farmers. (b) **Increased Dependency**: Reliance on pesticides can lead to reduced adoption of sustainable pest management practices, increasing long-term costs. (6) **Soil and Water Ecosystem Disruption**: (a) **Chemical Build-Up**: Repeated use can lead to pesticide accumulation in soil, affecting its quality and health. (b) **Harm to Aquatic Life**: Pesticides entering water bodies can harm fish, amphibians, and other aquatic organisms.

Fungicides: Fungicides, while useful for controlling fungal diseases in agriculture, can have several harmful effects on the environment, soil health, and human health when misused or overused. Here are the key concerns: (1) **Effects on Soil Health**: (a) **Soil Microbial Imbalance**: Fungicides can kill beneficial soil microorganisms, disrupting the natural balance and reducing soil fertility. (b) **Reduced Nutrient Cycling**: Harm to microbes involved in nutrient recycling can impair the soil's ability to supply nutrients to plants. (c) **Persistence in Soil**: Some fungicides have long half-lives, leading to accumulation and potential soil toxicity. (2) **Environmental Contamination**: (a) **Water Pollution**: Fungicides can leach into water bodies, contaminating groundwater, rivers, and lakes, harming aquatic ecosystems. (b) **Bioaccumulation**: Persistent fungicides can accumulate in the food chain, affecting higher organisms, including humans. (c) **Non-Target Impact**: They can harm non-target species like earthworms and beneficial fungi. (3) **Resistance Development**: (a) **Fungal Resistance**: Overuse or improper application can lead to resistant fungal strains, making fungicides less effective over time. (b) **Cross-Resistance**: Resistance to one fungicide may result in resistance to similar fungicides, reducing the options available for disease control. (4) **Impact on Biodiversity**: (a) **Harm to Beneficial Organisms**: Fungicides can negatively affect pollinators like bees, natural predators of pests, and mycorrhizal fungi, which are crucial for plant nutrient uptake. (b) **Ecosystem Imbalance**: Disruption in soil and plant ecosystems can have cascading effects on biodiversity. (5) **Human Health Risks**: (a) **Toxic Exposure**: Prolonged exposure to fungicides can lead to health issues, including skin irritation, respiratory problems, and endocrine disruption. (b) **Carcinogenic Potential**: Some fungicides are suspected or confirmed carcinogens. (c) **Residues in Food**: Improper fungicide use can leave residues on crops, posing risks to consumers. (6) Crop Damage: (a) **Phytotoxicity**: Some fungicides, especially when applied inappropriately, can damage crops, reducing yields and quality. (b) **Impact on Plant Growth**: Fungicides may interfere with beneficial fungi associated with plant roots, affecting growth and resilience.

Antibiotics:

The use of antibiotics in crop agriculture, while primarily intended for controlling bacterial diseases in plants, can have several harmful effects on the environment, human health, and the overall agricultural ecosystem. These effects are often overlooked, but they can lead to long-term consequences. (1) **Antibiotic Resistance**: (a) **Development of Resistance in Bacteria**: The overuse or misuse of antibiotics can lead to the development of antibiotic-resistant bacteria. These bacteria may become more difficult or even impossible to treat with conventional antibiotics, posing a serious public health threat. (b) **Transfer to Human Pathogens**: Resistant bacteria from plants or soil can transfer to human pathogens, rendering treatments for human infections less effective. (2) **Impact on Soil Health**: (a) **Disruption of Soil Microbiome**: Antibiotics can kill or inhibit beneficial soil microorganisms, such as nitrogen-fixing bacteria, fungi, and other microbes essential for nutrient cycling. (b) **Reduced Soil Fertility**: The disruption of beneficial microbes can impair soil health and reduce the soil's natural ability to support plant growth. (3) **Water Contamination**: (a) **Runoff into Waterways**: Antibiotics can leach into groundwater or be carried away by runoff into rivers, lakes, and streams, potentially contaminating drinking water sources and aquatic ecosystems. (b) **Harm to Aquatic Life**: The presence of antibiotics in water can harm aquatic organisms, including fish and invertebrates, disrupting ecosystems and biodiversity. (4) **Impact on Beneficial**

Insects: (a) **Harm to Pollinators**: Some studies suggest that antibiotics may negatively affect pollinators such as bees, which are essential for crop pollination. This can lead to declines in pollinator populations, impacting crop yields and biodiversity. (b) **Disruption of Pest Control**: Antibiotics may also harm beneficial insects that control pest populations, potentially leading to an increase in pest outbreaks. (5) Residue in Food: (a) **Health Risks from Residues**: Antibiotic residues on crops can enter the food supply and pose potential health risks to humans. Consuming foods with antibiotic residues can lead to allergic reactions, digestive disturbances, and the development of antibiotic resistance in human pathogens. (b) **Regulatory Challenges**: Many countries have stringent regulations regarding acceptable antibiotic residue levels in food, and violations can lead to market access issues and economic losses for farmers. (6) **Impact on Plant Health**: (a) **Toxicity to Plants**: Some antibiotics may cause phytotoxicity, damaging plant tissues and reducing crop yields. This can lead to poor plant growth and diminished harvest quality. (b) **Interference with Symbiosis**: Antibiotics can interfere with the symbiotic relationships between plants and beneficial microorganisms like mycorrhizal fungi, which are essential for nutrient uptake. (7) **Loss of Biodiversity**: (a) **Disruption of Microbial Communities**: The application of antibiotics can reduce microbial diversity in the soil, which is critical for maintaining ecosystem services such as nutrient cycling, pest control, and disease resistance. (b) **Reduction in Plant Diversity**: Antibiotics can reduce the growth of certain plant species in the soil, leading to a decline in overall plant biodiversity. (8) **Economic Costs**: (a) **Increased Resistance Management Costs**: The development of antibiotic resistance may require more expensive and less effective treatments, increasing costs for farmers. (b) **Market Access Issues**: Some international markets have strict regulations regarding antibiotic use in agriculture, and failure to comply with these regulations can limit a farm's access to global markets.

8.3.3 No irrigation:

While irrigation is essential for supporting crop growth in areas with inadequate rainfall, its misuse or overuse can lead to several harmful effects on the environment, soil health, and agriculture. Below are some key detrimental impacts of irrigation: (1) Soil Salinization: (a) **Salt Build-Up**: Irrigation, especially in arid and semi-arid regions, can lead to the accumulation of salts in the soil. As water evaporates, it leaves salts behind, which can degrade soil quality. (b) **Impact on Crop Growth**: High salinity levels reduce the ability of plants to take up water, leading to stunted growth, lower yields, and even plant death. (c) **Loss of Fertility**: Salinization reduces the soil's ability to retain nutrients, further affecting plant health. (2) **Waterlogging**: (a) **Excess Water in Soil**: When irrigation is poorly managed or applied excessively, the soil can become saturated with water, leading to waterlogging. (b) **Oxygen Deprivation**: Waterlogged soils reduce oxygen availability to plant roots, causing root rot and hindering nutrient uptake. (c) **Impact on Crop Health**: Crops like rice and certain legumes can tolerate waterlogging, but most crops such as wheat and corn are severely affected, leading to poor growth and reduced productivity. (3) **Decline in Groundwater Levels**: (a) **Over Extraction**: Excessive use of groundwater for irrigation can deplete aquifers, leading to a decline in water availability for both agriculture and human consumption. (b) **Subsidence**: In areas where groundwater is over-extracted, the land can sink or subside, leading to structural damage to infrastructure and further water management issues. (c) **Cost of Irrigation**: As groundwater levels decline, farmers may need to drill deeper wells, increasing costs and reducing the sustainability of irrigation practices. (4) **Soil Erosion**: (a) **Surface Runoff**: Poor irrigation practices can lead to water runoff, which may carry away topsoil, particularly in areas with sloped terrain or inadequate drainage. (b) **Loss of Fertile Topsoil**: Erosion removes the nutrient-rich upper layers of the soil, reducing its fertility and increasing the need for fertilizers. (c) **Siltation:** Runoff can carry sediment into water bodies, causing siltation and affecting water quality and aquatic ecosystems. (5) **Nutrient Leaching:** (a) **Excessive Water Use:** Over-irrigation can cause the leaching of valuable soil nutrients like nitrogen, phosphorus, and potassium beyond the root zone, making them unavailable to plants. (b) **Pollution of Water Bodies:** Leached nutrients can run off into nearby rivers, lakes, and groundwater, leading to nutrient pollution and potentially contributing to harmful algal blooms. (6) **Aquatic Ecosystem Damage:** (a) **Diversion of Water:** Irrigation often involves the diversion of water from rivers, lakes, and streams, disrupting natural water flows and harming aquatic ecosystems. (b) **Decreased Biodiversity:** Reduced water availability for aquatic habitats can lead to a loss of biodiversity, affecting fish, plants, and other organisms. (c) **Contamination:** Irrigation runoff containing pesticides, fertilizers, and salts can pollute water bodies, further harming aquatic life. (7) **Increased Pest and Disease Pressure:** (a) **Moisture Favoring Pests:** Stagnant water and excessive moisture in irrigated fields create a favorable environment for pests like mosquitoes and diseases such as fungal infections. (b) **Spread of Pathogens:** Over-irrigation can promote the spread of plant diseases, such as root rot and fungal pathogens, which thrive in wet conditions. (8) **Irrigation-Induced Climate Change Effects:** (a) **Greenhouse Gas Emissions:** Irrigated fields, particularly those that are waterlogged, can produce methane—a potent greenhouse gas—especially in rice paddies. (b) **Increased Carbon Footprint:** The energy used for irrigation (e.g., pumping groundwater or running irrigation systems) contributes to greenhouse gas emissions, increasing the carbon footprint of agriculture. (9) **Land Degradation and Desertification:** (a) **Loss of Productive Land:** Over-irrigation, particularly in arid regions, can lead to desertification, where previously productive land becomes barren due to salinization and soil degradation.

(b) **Increased Vulnerability to Drought:** Over-reliance on irrigation can make agricultural systems more vulnerable to water shortages and droughts when irrigation infrastructure fails or when water sources become depleted.

8.3.4 Mulching:

Mulching is a layer of natural organic materials such as dry leaves which is intentionally maintained permanently on the soil surface, to improve soil health. Organic mulching offers numerous benefits for ecological agriculture, supporting soil health, plant growth, and environmental sustainability. Here are some of the key advantages: (1) **Soil Moisture Retention:** (a) **Reduced Evaporation:** Organic mulches, such as straw, leaves, or grass clippings, help retain moisture in the soil by reducing evaporation. This is particularly beneficial during hot and dry conditions. (b) **Consistent Moisture Levels:** Mulch helps maintain more stable soil moisture, reducing the risk of drought stress and supporting consistent plant growth. (2) **Weed Suppression:** (a) **Competition Reduction:** By blocking sunlight, organic mulches prevent weed seeds from germinating and growing, reducing the need for chemical herbicides. (b) **Decreased Weed Pressure:** Less weeding is required, saving time and labor, while promoting healthier crop growth. (3) **Soil Temperature Regulation:** (a) **Cooler Roots in Summer:** Mulch helps keep the soil cooler during hot weather, protecting plant roots from extreme heat and reducing the risk of heat stress. (b) **Warmer Soil in Winter:** In colder months, mulch acts as an insulating layer, helping to protect roots from freezing and promoting early plant growth in the spring. (4) **Improved Soil Structure and Health:** (a) **Organic Matter Addition:** As organic mulches break down, they add valuable organic matter to the soil, improving soil structure and fertility over time. (b) **Enhanced Soil Microbial Activity:** Organic mulches support beneficial soil microorganisms by providing a habitat and food source, boosting soil biodiversity and nutrient cycling. (5) **Erosion Control:** (a) **Protection from Water Erosion:** Mulch helps reduce the impact of rainfall on the soil, minimizing the risk of water erosion by protecting the soil surface. (b) **Wind Erosion Mitigation:** Mulch acts as a barrier, protecting the soil from wind erosion, especially in areas prone to high winds. (6.) Improved Soil Fertility: (a) **Nutrient Recycling:** As organic mulches decompose, they release essential nutrients such as nitrogen, phosphorus, and potassium into the soil, enriching soil fertility. (b) **Balanced Soil pH:** Organic mulches can help buffer the soil pH, preventing it from becoming too acidic or alkaline, which is beneficial for plant growth. (7) **Pest and Disease Management:** (a) **Reduction of Soil-borne Diseases:** Mulching can help protect plants from soil-borne diseases by creating a physical barrier between the soil and plant stems, reducing the spread of pathogens. (b) **Habitat for Beneficial Insects:** Organic mulch provides shelter for beneficial insects such as predatory beetles, which help control pests. (8) **Reduced Soil Compaction:** Protection from Foot Traffic: Mulch helps protect the soil from compaction caused by foot traffic or machinery, maintaining soil structure and promoting healthy root development. (b) Improved Water Infiltration: By preventing compaction, mulch allows water to infiltrate the soil more easily, reducing runoff and improving water absorption. (9) **Sustainable and Low-Cost:** (a) **Renewable Resource:** Organic mulching materials, such as grass clippings, leaves, and straw, are renewable and often readily available on-site or locally sourced, making it a cost-effective option. **Reduction in Synthetic Inputs:** By suppressing weeds and enriching soil fertility, mulching reduces the need for synthetic fertilizers, herbicides, and pesticides, aligning with the principles of ecological agriculture. (10) **Carbon Sequestration: (a)** Reduction in Carbon Emissions: Organic mulches help sequester carbon in the soil as they break down, contributing to carbon storage and reducing greenhouse gas emissions. (b) **Improved Soil Carbon Stocks:** Over time, organic mulching can increase the amount of organic carbon in the soil, enhancing its long-term sustainability and supporting climate mitigation. (11) **Aesthetic and Functional Benefits:** (a) **Enhanced Visual Appeal:** Mulching improves the appearance of garden beds and agricultural plots, contributing to the aesthetic value of a farm or garden. (b) **Reduced Soil Splashing:** Mulch helps prevent soil splashing onto plants during rain, reducing the spread of soil-borne diseases.

8.3.5 Cover cropping:

Cover cropping is a sustainable agricultural practice where plants are grown primarily to improve soil health, protect the soil, and enhance farming ecosystems. It offers several benefits for ecological agriculture: (1) **Soil Health Improvement:** (a) **Organic Matter Addition:** Cover crops increase soil organic matter as they decompose, improving soil structure, fertility, and water-holding capacity. (b) **Soil Microbial Activity:** They support soil microbes by providing food and habitat, promoting nutrient cycling and soil biodiversity. (2) **Erosion Control:** (a) **Protection from Wind and Water:** Cover crops protect the soil from erosion by stabilizing it with their roots and reducing the impact of rainfall and wind. (b) **Reduced Runoff:** Their root systems help water infiltrate the soil, reducing surface runoff and preventing the loss of topsoil. (3) **Nutrient Management:** (a) **Nitrogen Fixation:** Leguminous cover crops (e.g., clover, vetch) fix atmospheric nitrogen, enriching the soil for subsequent crops. (b) **Nutrient Scavenging:** Deep-rooted cover crops capture nutrients like nitrogen and phosphorus that might otherwise leach out of the soil, making them available for future crops. (4) **Weed Suppression:** (a) **Competition with Weeds:** Cover crops outcompete weeds for sunlight, water, and nutrients, reducing weed pressure without the need for herbicides. (b) **Allelopathic Effects:** Some cover crops (e.g., rye) release chemicals that inhibit weed seed germination. (5) **Soil Moisture**

Retention: (a) **Reduced Evaporation:** Cover crops help retain soil moisture by reducing evaporation, particularly in dry climates. (b) **Improved Water Infiltration:** Their root systems enhance soil porosity, allowing better water penetration and storage. (6) **Pest and Disease Management:** (a) **Habitat for Beneficial Insects:** Cover crops provide shelter and food for beneficial insects, such as pollinators and predators of pests. (b) **Breaking Pest Cycles:** By interrupting the life cycles of pests and pathogens, they reduce infestations and soil-borne diseases. (7) **Improved Soil Structure:** (a) **Reduced Compaction:** Deep-rooted cover crops, such as radishes, break up compacted soil layers, improving root penetration and aeration. (b) **Aggregation of Soil Particles:** Their organic matter binds soil particles together, creating a crumbly, fertile structure ideal for plant growth. (8) **Biodiversity Enhancement:** (a) **Increased Plant Diversity:** Cover crops enhance on-farm biodiversity, supporting a wider range of organisms above and below ground. (b) **Ecosystem Stability:** They contribute to ecological balance by fostering diverse biological communities. (9) Climate Resilience: (a) **Carbon Sequestration:** Cover crops capture atmospheric carbon and store it in the soil, helping mitigate climate change. (b) **Improved Drought Tolerance:** Their water-retention benefits help soils withstand drought conditions more effectively. (10) **Economic Benefits:** (a) **Reduced Input Costs:** By fixing nitrogen and suppressing weeds, cover crops reduce the need for synthetic fertilizers and herbicides. (b) **Increased Crop Yields:** Over time, healthier soils and reduced weed competition lead to higher productivity. (11) **Water Quality Protection:** (a) **Reduced Nutrient Runoff:** By capturing excess nutrients, cover crops prevent pollution of waterways and reduce the risk of eutrophication. (b) **Improved Groundwater Quality:** They minimize the leaching of nitrates into groundwater, protecting drinking water supplies. (12) **Soil Protection in Off-Season:** Cover crops keep the soil covered during fallow periods, preventing degradation and maintaining productivity.

8.3.6 Multilayer Polycropping:

A multilayer polycropping system, where crops of different heights, root structures, and growth cycles are cultivated together, offers numerous benefits to soil health by mimicking natural ecosystems. Here are the key advantages: (1) **Soil Erosion Control:** (a) **Continuous Ground Cover:** The dense vegetation and ground cover provided by multilayer crops protect the soil from wind and water erosion. (b) **Root Systems as Anchors:** The varied root depths stabilize the soil, preventing it from washing or blowing away. (2) **Enhanced Soil Organic Matter:** (a) **Leaf Litter Contribution:** Falling leaves and crop residues from different layers add organic matter to the soil, enriching it over time. (b) **Decomposition Cycle:** Organic matter decomposition improves soil fertility and promotes microbial activity. (3) **Nutrient Cycling and Availability:** (a) **Efficient Nutrient Use:** Crops with different root depths access nutrients at various soil levels, reducing nutrient competition and depletion. (b) **Nutrient Recycling:** Organic matter returned to the soil through mulch, crop residues, and root decay enhances nutrient availability. (4) **Improved Soil Structure:** (a) **Diverse Root Systems:** Varied root structures (deep, shallow, fibrous, and taproots) enhance soil porosity and aggregate stability, creating a better environment for water and air movement. (b) **Reduced Compaction:** Continuous crop cover and root activity prevent soil compaction and crusting. (5) **Increased Soil Microbial Activity:** (a) **Microbial Habitat:** Organic matter and root exudates from diverse crops feed soil microorganisms, boosting microbial diversity and activity. (b) **Symbiotic Relationships:** Leguminous crops in the system foster nitrogen-fixing bacteria, enhancing soil nitrogen levels. (6) **Weed Suppression:** (a) **Dense Canopy:** Multilayer crops reduce sunlight reaching the soil surface, inhibiting weed germination and growth. (b) **Natural Mulching:** Crop residues act as a mulch layer, further suppressing weeds and conserving soil health. (7) **Soil Moisture Conservation:** (a) **Reduced Evaporation:** The canopy layers shade the soil, minimizing water loss through evaporation. (b) **Improved Water Infiltration:** Root channels enhance water infiltration, reducing runoff and improving water availability for crops. (8) **Carbon Sequestration:** (a) **Soil Carbon Storage:** The addition of organic matter from multiple crops increases soil organic carbon, improving long-term soil fertility and mitigating climate change. (b) **Root Carbon Input:** Continuous root activity contributes to carbon storage in deeper soil layers. [9] **Pest and Disease Resistance:** (a) **Break in Pest Cycles:** Diverse crops interrupt pest and pathogen life cycles, reducing the build-up of soil-borne diseases. (b) **Beneficial Microorganisms:** Increased microbial diversity helps suppress soil pathogens and maintain soil health. (10) **Improved Nutrient Use Efficiency:** (a) **Complementary Nutrient Uptake:** Different crops have unique nutrient requirements, reducing nutrient competition and promoting balanced soil fertility. (b) **Leguminous Crops:** Inclusion of legumes fixes atmospheric nitrogen, enriching the soil for other crops in the system. (11) **Sustainable Soil Fertility:** (a) **No Dependence on External Inputs:** By cycling nutrients naturally, multilayer systems reduce the need for chemical fertilizers, enhancing long-term soil health. (b) **Natural Fertilization:** Organic residues contribute to the slow release of nutrients, ensuring sustainable fertility management. (12) **Protection Against Soil Degradation:** (a) **Minimized Land Disturbance:** The continuous cover and diverse cropping reduce soil degradation and improve resilience to environmental stressors. (b) **Regeneration of Degraded Soils:** Organic inputs and microbial activity help rehabilitate poor-quality soils. (13) **Biodiversity and Ecosystem Balance:** (a) **Encouragement of Soil Fauna:** Earthworms, insects, and other soil organisms thrive in the organic-rich, diverse habitat, contributing to soil aeration and nutrient cycling. (b) **Resilience to Climate Stress:** A diverse system is better equipped to adapt to temperature fluctuations and water scarcity, preserving soil integrity. **Example of a Multilayer Polycropping System:** (a) **Top Layer:** Coconut, areca nut, or tall fruit trees like mango or jackfruit. (b) **Middle Layer:** Bananas, papayas, or shrubs. (c) **Ground Layer:** Vegetables like spinach, beans, or herbs. (d) **Underground Layer:** Root crops like turmeric, ginger, or sweet potatoes.

8.3.7 Probiotics:

The use of home-made crude microbial cultures (probiotics) in traditional agriculture is a time-tested practice that promotes soil health, plant growth, and sustainable farming. These microbial cultures consist of beneficial microorganisms such as bacteria, fungi, and yeasts that improve nutrient availability, suppress pathogens, and enhance soil fertility. Traditional farmers often prepare these cultures using simple, cost-effective, and natural ingredients available locally. **Common Home–Made Microbial Cultures in Traditional Agriculture:** (1) **Jeevamrut (India):** Ingredients: Cow dung, cow urine, jaggery, gram flour, and water. Process: Fermented for 3–5 days to encourage microbial growth. Use: Applied to the soil or as a foliar spray to enhance microbial activity and plant nutrition. (2) **Fermented Plant Juice (Korea):** Ingredients: Fresh plant leaves, brown sugar, and water. Process: Fermented for 7–10 days. Use: Provides a rich source of nutrients and beneficial microbes for plants. (3) **Effective Microorganisms (EM):** Ingredients: Lactic acid bacteria, yeast, and photosynthetic bacteria mixed with molasses and water. Process: Fermented anaerobically for 7–10 days. Use: Improves soil health and plant resilience. (4] **Fish Amino Acid (Japan):** Ingredients: Fish waste, jaggery or sugar, and water. Process: Fermented for 1–2 weeks. Use: Provides nitrogen and other nutrients while enhancing microbial diversity. (5) **Cow Pat Pit (CPP):** Ingredients: Cow dung, crushed eggshells, basalt powder, and fermented in a pit. Process: Fermented for 3–4 months. Use: Boosts soil fertility and microbial activity.

8.3.8 Mixed animal husbandry (integrated farming):

Mixed animal husbandry involves raising diverse animal species together, such as cattle, goats, sheep, poultry, and pigs, within a farming system. This practice can significantly enhance soil health through complementary ecological processes. Here are the key benefits: (1) **Enhanced Organic Matter Addition:** (a) **Diverse Manure Sources:** Different animals produce manure with varying nutrient compositions. This enriches the soil with a wider range of nutrients, improving fertility and microbial activity. (b) **Decomposition and Mulching:** Manure and bedding materials decompose, adding organic matter that enhances soil structure and water retention. (2) **Improved Nutrient Cycling:** (a) **Efficient Recycling:** Animals convert crop residues and by-products into manure, which can be applied to fields, closing the nutrient loop. (b) **Varied Grazing Patterns:** Different grazing behaviors ensure even nutrient distribution and prevent nutrient hotspots. (3) **Soil Microbial Diversity:** (a) **Manure as a Microbial Source:** Animal manure introduces beneficial microorganisms that enhance soil microbial diversity and nutrient cycling. (b) **Reduced Pathogen Pressure:** Mixed species reduce the risk of specific pathogens dominating, maintaining a balanced soil ecosystem. (4) **Soil Structure Improvement:** (a) **Natural Aeration:** Grazing animals gently disturb the soil surface, improving aeration without causing compaction when managed properly. (b) **Organic Carbon Addition:** Organic matter from manure improves soil aggregation, making the soil crumbly and resistant to erosion. (5) **Weed Suppression:** (a) **Targeted Grazing:** Animals like goats and sheep consume weeds, reducing weed pressure without the need for herbicides. (b) **Natural Mulching:** Trampling by animals can help create a natural mulch layer that suppresses weed growth.

(6) **Soil Moisture Retention:** (a) **Improved Water Holding Capacity:** The organic matter from animal manure and bedding increases the soil's ability to retain water, reducing drought stress. (b) **Reduced Runoff:** Mixed husbandry systems with ground cover help reduce water runoff, conserving soil moisture. (7) **Pest and Disease Management:** (a) **Breaking Pest Cycles:** By integrating multiple species, pests and pathogens associated with a single type of livestock are less likely to thrive. (b) **Predator Attraction:** Mixed animal systems can attract beneficial predators, such as birds, which help control soil pests. (8) **Erosion Control:** (a) **Improved Vegetative Cover:** Livestock systems often include pasture management, which maintains soil cover and prevents erosion. (b) **Manure as a Binding Agent:** Organic matter from manure stabilizes soil particles, reducing susceptibility to erosion. (9) **Carbon Sequestration:** (a) **Manure-Driven Carbon Storage:** Regular application of manure and compost enhances soil organic carbon, contributing to long-term carbon storage. (b) **Perennial Pastures:** If mixed husbandry is paired with rotational grazing, it promotes perennial grass systems that sequester atmospheric carbon in the soil. (10) **Diverse Grazing Benefits:** (a) **Species-Specific Grazing Patterns:** Different animals graze on different plants and layers of vegetation, ensuring that no single plant dominates and promoting plant diversity. (b) **Root System Stimulation:** Grazing encourages plant regrowth, stimulating deeper and healthier root systems that improve soil stability and fertility. (11) **Enhanced Soil Fertility:** (a) **Balanced Nutrient Distribution:** The presence of multiple species ensures that nutrients are distributed evenly, as different animals prefer different areas for grazing and resting. (b) **Improved Mineral Availability:** Manure enhances the availability of essential minerals such as nitrogen, phosphorus, and potassium. (12) **Reduced Soil Degradation:** (a) **Sustainable Grazing:** Properly managed mixed husbandry reduces overgrazing and soil compaction, mitigating soil degradation. (b) **Regeneration**

of Marginal Lands: Animal manure and rotational grazing can rehabilitate degraded lands by improving soil organic matter and fertility. (13) **Integrated Farming Synergies:** (a) **Crop–Livestock Integration:** Residues from crops can be used as feed, while manure from livestock returns to the fields, creating a symbiotic system. (b) **Eliminating Input Costs:** Mixed animal husbandry minimizes the need for synthetic fertilizers and herbicides by naturally enriching the soil.

8.4 Interspecific Interactions in Soil Health Management:

Interspecific interactions between plant roots, earthworms and microorganisms play a crucial role in maintaining soil health through various mechanisms.

8.4.1 Earthworms – the Ecosystem Engineer:

Earthworms are nature's tillers (bioturbation) and recyclers, playing an indispensable role in creating healthy, fertile soils. Earthworms play a crucial role in maintaining and improving soil health, making them essential for sustainable agriculture and natural farming practices. Here are the key contributions of earthworms to soil health management: (1) **Soil Aeration and Structure Improvement: Burrowing:** Earthworms create tunnels as they move through the soil, improving aeration and water infiltration. **Soil Aggregation:** Their movement helps bind soil particles into aggregates, improving soil structure and reducing erosion. (2) **Nutrient Cycling and Availability: Decomposition of Organic Matter:** Earthworms consume organic materials like plant residues, breaking them down into simpler compounds. **Cast Production:** Their excreta (casts) are rich in nutrients like nitrogen, phosphorus, and potassium, making these nutrients readily available to plants. **Enhancing Microbial Activity:** Earthworms stimulate microbial populations, which play a vital role in nutrient mineralization. (3) **Humus Formation:** Earthworms contribute to the formation of humus by breaking down organic matter into stable organic compounds, improving soil fertility and water retention. (4) **Water Retention:** The burrows created by earthworms improve soil porosity, enabling better water retention and distribution. This reduces waterlogging in heavy soils and enhances water availability in drier conditions. (5) **pH Regulation:** Earthworm casts can help buffer soil pH, making the soil more suitable for plant growth. (6) **Biological Pest Control:** Earthworms help suppress soil-borne pests by promoting beneficial microbes that compete with or inhibit harmful organisms. (7) **Carbon Sequestration:** Earthworms aid in stabilizing organic carbon in the soil, contributing to carbon sequestration and reducing greenhouse gas emissions. (8) **Vermicomposting:** Earthworms are used in vermicomposting to convert organic waste into nutrient-rich vermicompost, which improves soil fertility. Vermicompost enhances plant growth, soil structure, and microbial activity. (9) **Soil Detoxification:** Earthworms can help break down and immobilize pollutants in the soil, such as heavy metals and pesticides, reducing their bioavailability and toxicity. (10) **Enhanced Plant Growth:** Soils with active earthworm populations have higher nutrient availability, better structure, and improved aeration, directly supporting robust plant growth.

8.4.2 Arbuscular Mycorrhizal Fungi (AMF):

Mutualism between arbuscular mycorrhizal fungi (AMF) and plant roots is a symbiotic relationship where both organisms benefit. This association is one of the most ancient and widespread mutualisms in terrestrial ecosystems, involving about 80% of vascular plants and members of the Glomeromycota fungal phylum. (1) **Fungal Role: Nutrient Acquisition:** AMF extend their hyphae into the soil, absorbing nutrients (especially phosphorus, nitrogen, and micronutrients) that are otherwise inaccessible to plant roots. **Soil**

Structure: Their hyphal network improves soil aggregation and stability. **Protection Against Stress**: AMF protect plants from pathogens, drought, and heavy metal toxicity. (2) **Plant Role: Carbon Supply**: Plants provide AMF with carbohydrates (mainly glucose and sucrose) derived from photosynthesis, as fungi cannot photosynthesize. **Habitat**: The plant root system offers a secure environment for fungal colonization.

Mechanism of Interaction: (1) **Colonization**: Fungal spores germinate and grow toward plant roots, forming a physical and chemical connection. Fungi penetrate the root cortex, establishing structures like arbuscules (site of nutrient exchange) and vesicles (storage). (2) **Nutrient Exchange**: The arbuscules create a large surface area for bidirectional transfer of nutrients and sugars. Plants absorb nutrients transported by AMF, while fungi receive sugars from plant photosynthesis. (3) **Hyphal Network**: AMF hyphae extend beyond the root zone, exploring a larger soil volume for nutrient uptake. This "mycorrhizosphere" also enhances microbial activity and nutrient cycling.

Ecological and Agricultural Importance: (1) **Enhanced Nutrient Uptake**: AMF improve the bioavailability of phosphorus and nitrogen, crucial for plant growth. Particularly beneficial in nutrient-poor or degraded soils. (2) **Drought Resistance**: Hyphae increase water absorption and retention, helping plants withstand drought conditions. (3) **Disease Resistance**: AMF suppress root pathogens by outcompeting them or inducing plant defense mechanisms. (4) **Improved Soil Structure**: Fungal hyphae produce glomalin, a glycoprotein that strengthens soil aggregates and enhances carbon sequestration. (5) **Eliminate Dependency on Fertilizers**: AMF can eliminate the need for chemical fertilizers, promoting sustainable agriculture. (6) **Biodiversity**: Supports a healthy soil ecosystem by interacting with other beneficial microbes and plants.

8.4.3 Plant Growth Promoting Rhizobacteria (PGPR):

The mutualism between plant growth promoting rhizobacteria (PGPR) and plant roots is a symbiotic relationship where both the bacteria and the plants benefit. PGPR are beneficial bacteria that colonize the rhizosphere (the root-soil interface) or plant roots and enhance plant growth through various direct and indirect mechanisms. (1) **Role of PGPR: Nutrient Acquisition:** PGPR facilitate the availability of nutrients like nitrogen, phosphorus, and essential micronutrients. **Stress Tolerance:** Help plants withstand biotic (pathogens) and abiotic (drought, salinity) stresses. **Growth Promotion:** Produce phytohormones and enhance root architecture for better nutrient and water uptake. (2) **Role of Plants: Carbon Supply:** Provide PGPR with exudates like sugars, amino acids, and organic acids from the roots, serving as a nutrient source. **Habitat:** Offer a stable environment in the rhizosphere for bacterial colonization.

Mechanisms of Interaction: (1) **Direct Plant Growth Promotion: Nitrogen Fixation:** Some PGPR (e.g., Rhizobium, Azospirillum) fix atmospheric nitrogen into forms usable by plants. **Phosphorus Solubilization:** Solubilize insoluble phosphates into plant-accessible forms. **Production of Phytohormones:** Auxins (e.g., Indole-3-acetic acid) enhance root elongation and branching. Cytokinins and gibberellins promote cell division and shoot growth. **Iron Acquisition:** Secrete siderophores that chelate iron, making it more available to plants. (2) **Indirect Plant Growth Promotion: Disease Suppression:** Compete with pathogenic microorganisms for space and nutrients in the rhizosphere. Produce antimicrobial substances to inhibit pathogen growth. **Induced Systemic Resistance (ISR):** Trigger plant immune responses to better resist pests and diseases. (3) **Root Colonization:** PGPR adhere to root surfaces or invade the root tissue, forming biofilms that ensure their stability and function in the rhizosphere.

Examples of PGPR and Their Functions: (1) **Nitrogen-Fixing Bacteria: Rhizobium:** Forms nodules on legume roots, fixing nitrogen symbiotically. **Azospirillum:** Associative nitrogen fixation in cereals like wheat and rice. (2) **Phosphate-Solubilizing Bacteria: Pseudomonas fluorescens:** Solubilizes inorganic phosphates. **Bacillus megaterium:** Releases organic acids to mobilize phosphorus. (3) **Phytohormone-Producing Bacteria: Azotobacter:** Produces auxins and gibberellins to enhance root and shoot growth. (4) **Stress Alleviating Bacteria: Bacillus subtilis:** Produces enzymes that break down harmful reactive oxygen species (ROS) during stress. **Pseudomonas putida:** Enhances drought tolerance by improving root water uptake.

8.5 Biomonitoring of soil health:

Soil health is the health conditions of the living organisms in soil such as plants, animals and microorganisms, therefore soil health cannot be tested in the laboratory by analysing dead soil samples. Soil health must be monitored in the field continuously by visually observing the health conditions of the living organisms in a given soil. The most convenient way to monitor soil health is to study the indicator plants (such as sunflower), indicator animals (earthworms) and indicator microorganisms (ectomycorrhizal mushrooms).

8.5.1 Biomonitoring of Plant Health:

Plant health is usually determined by the intensity of green colour of the leaves as well as the absence of damage or disease symptoms in leaves (Debarat Bhaduri et al, 2022). (See 9.4).

8.5.2 Biomonitoring of Animal Health:

Animal health is usually determined by the abundance, species richness, vitality and absence of disease symptoms or weakness in earthworms (Heinz-Christian Frund et al., 2010).

8.5.3 Biomonitoring of Microorganism Health:

The health conditions of the soil microorganisms is determined by the species richness and normal physical appearance of the wild mushroom species naturally occuring in a soil (Sashika D, 2023).

CHAPTER 9

PLANT HEALTH MANAGEMENT

Plant health is the capacity of the plants to sustain normal growth, development and reproduction without the help of any external inputs or irrigation, and absence of any symptoms of damage or disease in the plant bodies.

9.1 Objectives of Plant Health Management:

The objectives of plant health management in natural farming includes (1) Automation of ecosystem services, (2) Conservation of natural capital (natural resources), (3) Phytoremediation of environmental pollution, (4) Improvement of human health, (5) Minimising risk and cost of production of crops, (6) Maximising, aggregate yield, product quality, product price and maximising profit of the farmers.

9.1.1 Automation of Ecosystem Services:

Ecosystem services are the benefits that humans derive from healthy ecosystems, essential for survival and quality of life. They are categorized into four main types: **Provisioning Services:** These include food, water, raw materials, and medicinal resources. **Regulating Services:** These services control climate, air quality, and disease, and include processes like pollination and flood regulation. **Supporting Services:** These underpin other services by maintaining biodiversity and nutrient cycling. **Cultural Services:** These encompass recreational, aesthetic, and spiritual benefits derived from nature. Here are some important ecosystem services in natural farming:

Nitrogen Cycle in Natural Farming Agroecosystem: The nitrogen cycle in an agroecosystem is a crucial process that ensures the availability of nitrogen, an essential nutrient for plant growth. It involves the movement of nitrogen through various forms and components, including the soil, plants, animals, microorganisms, and the atmosphere. Here's an overview of the nitrogen cycle in an agroecosystem: (1) **Nitrogen Fixation**: Atmospheric nitrogen (N_2) is converted into ammonia (NH_3) or related compounds that plants can absorb and utilize. **Sources of Nitrogen Fixation: Biological Fixation**: Symbiotic bacteria (e.g., Rhizobium in legumes) or free-living bacteria (e.g., Azotobacter, Cyanobacteria). **Lightning**: Converts atmospheric nitrogen into nitrates that are deposited in the soil via rain. (2) **Ammonification (Decomposition)**: Organic nitrogen in dead plants, animals, and animal waste is converted into ammonium (NH_4^+) by decomposer microorganisms. Key Role: Recycles nitrogen within the agroecosystem. (3) **Nitrification**: Ammonium (NH_4^+) is converted into nitrites (NO_2^-) and then into nitrates (NO_3^-) by nitrifying bacteria. Nitrosomonas: Converts NH_4^+ to NO_2^-. Nitrobacter: Converts NO_2^- to NO_3^-. Nitrates (NO_3^-) are the primary form of nitrogen absorbed by plants. (4] **Plant Uptake**: Plants absorb ammonium (NH_4^+) and nitrate (NO_3^-) from the soil to synthesize proteins, enzymes, and other essential compounds. (5) **Consumption by Animals**: Animals obtain nitrogen by consuming plants. The nitrogen is used to build animal proteins and other compounds. Animal waste returns nitrogen to the soil. (6) **Denitrification**: Nitrates (NO_3^-) are converted back into nitrogen gas (N_2) or nitrous oxide (N_2O) by denitrifying bacteria (e.g., Pseudomonas). Key Role: Releases nitrogen back into the atmosphere, completing the cycle.

Phosphorus Cycle in Natural Farming Agroecosystem: The phosphorus cycle in an agroecosystem is critical for maintaining soil fertility and ensuring plant growth, as phosphorus is an essential nutrient for energy transfer, photosynthesis, and root development in plants. Unlike the nitrogen cycle, the phosphorus cycle does not involve a gaseous phase and is primarily a sedimentary cycle. Here's an overview of how phosphorus moves through an agroecosystem: (1) **Weathering of Rocks**: Phosphorus originates from phosphate minerals in rocks. Weathering releases phosphate ions (PO_4^{3-}) into the soil, making them available for plants to absorb. This process is slow and depends on natural erosion and chemical reactions. (2) **Absorption by Plants**: Plants absorb phosphorus from the soil in the form of soluble phosphate ions (PO_4^{3-} and $H_2PO_4^-$). Phosphorus is used in the synthesis of DNA, RNA, ATP, and other essential molecules. (3) **Transfer to Animals**: When animals consume plants, phosphorus moves into the animal system, where it is used in bones, teeth, and metabolic processes. Phosphorus returns to the soil through animal waste and decomposition. (4) **Decomposition of Organic Matter**: Dead plants, animals, and organic waste are decomposed by microorganisms, releasing phosphorus back into the soil as inorganic phosphates. (5) **Formation of Soil Phosphate Pools**: Phosphorus in the soil exists in three forms: (a) **Available Phosphates**: Immediately usable by plants. (b) **Fixed Phosphates**: Bound to soil particles, not readily available. (c) **Organic Phosphates**: Part of organic matter, requiring decomposition to become available. (6) **Leaching and Runoff**: Excess phosphorus from soil and fertilizers can be lost to waterways through leaching and surface runoff. This contributes to eutrophication, where nutrient accumulation in water bodies leads to algal blooms and oxygen depletion. (7) **Sedimentation**: Phosphorus that reaches water bodies settles into sediments and becomes part of the long-term phosphorus pool. Over time, geological processes may uplift these sediments, starting the cycle anew.

Potassium Cycle in Natural Farming Agroecosystem: The potassium cycle in an agroecosystem describes the movement of potassium (K), an essential macronutrient for plant growth, within soil, plants, and the surrounding environment. Potassium is crucial for plant physiological processes, such as enzyme activation, photosynthesis, water regulation, and disease resistance. Unlike nitrogen, potassium does not form gaseous compounds and cycles primarily through soil and plant systems. (1) **Soil Potassium Pools:** (a) **Unavailable Potassium:** Found in primary minerals (e.g., feldspar, mica). Locked in crystalline structures and not immediately available to plants. Becomes available through slow weathering over geological time. (b) **Slowly Available Potassium:** Held in secondary minerals and interlayer spaces of clay particles. Released gradually through weathering and microbial activity. (c) **Readily Available Potassium:** Dissolved in soil water as potassium ions (K^+). This is the form directly absorbed by plant roots. (2) **Weathering of Minerals:** Potassium is released from primary minerals like feldspar and mica through weathering processes. This release is slow and contributes to the long-term potassium supply in soils. (3) **Plant Uptake:** Plants absorb potassium from the soil solution as K^+ ions. Potassium is essential for: Regulating water movement and stomatal function. Enhancing enzyme activity and photosynthesis. Improving crop quality and resistance to stress. (4) **Return to Soil through Crop Residues and Manure:** When plants die or are harvested, some potassium returns to the soil through crop residues, decomposing roots, or animal manure. (5) **Leaching:** Potassium, being water-soluble, can be lost from the root zone through leaching, especially in sandy or light soils. This is a significant loss pathway in high rainfall areas. (6) **Fixation in Clay:** Some potassium becomes fixed in clay minerals, such as illite or vermiculite, and is temporarily unavailable for plant uptake. Over time, this fixed potassium may become available again.

Pollination: Biotic pollination is the process where pollen is transferred from the male part (anther) of a flower to the female part (stigma) by living organisms, primarily insects and birds. This type of pollination includes **entomophily,** where insects like bees and butterflies are the main pollinators, and **ornithophily**, involving birds. Approximately 80% of plant pollination is biotic, highlighting its critical role in plant reproduction and biodiversity.

Natural Control of Insect Pests: Interspecific interactions play a crucial role in the control of insect pests in natural farming agroecosystems. These interactions, involving various organisms like predators, parasitoids, plants, and microbes, help maintain pest populations at manageable levels without relying on chemical pesticides. Key types of interspecific interactions include: (1) **Predation**: Predators feed on insect pests, reducing their populations. Examples: Lady Beetles (Coccinellidae): Feed on aphids, mealybugs, and whiteflies. Spiders: Consume a wide variety of insect pests like leafhoppers and moths. Hoverfly Larvae: Prey on aphids and small soft-bodied pests. (2) **Parasitism**: Parasitoids live on or inside pest organisms, eventually killing them. Examples: Braconid Wasps: Parasitize caterpillars and aphids, laying eggs inside them. Trichogramma spp.: Parasitize eggs of lepidopteran pests like borers and armyworms. Tachinid Flies: Lay eggs on pests like stink bugs and beetles, whose larvae feed on and kill the pests. (3) **Mutualism**: Both species benefit from the interaction, often indirectly suppressing pest populations. Examples: Ants and Plants: Some plants (e.g., Acacia) provide shelter and food (nectaries) for ants, which protect the plants by preying on herbivorous pests. Pollinators and Plants: Pollinators, while benefiting from nectar, contribute to healthy plant growth, indirectly reducing pest susceptibility. (4) **Competition**: Beneficial organisms or other pest species compete with harmful pests for resources, suppressing their populations. Examples: Soil Microbes: Beneficial microbes compete with pest-causing pathogens in the soil, reducing pest infestations. Predatory Insects: Compete with pest species for habitat and food. (5) **Commensalism**: One species benefits without affecting the other, often providing pest control indirectly. Examples: Epiphytic microorganisms on plants outcompete harmful pathogens, indirectly controlling pest populations. (6) **Allelopathy**: Certain plants release chemicals that deter or kill insect pests. Examples: Marigold (Tagetes spp.): Produces allelopathic compounds that repel nematodes and soil-borne pests. Neem (Azadirachta indica): Releases azadirachtin, which acts as an insect repellent and disrupts pest reproduction. (7) **Biological Control Agents**: Role in Interactions: Natural predators, parasites, and pathogens interact with pests to control their populations. Examples: Entomopathogenic Fungi: Fungi like Beauveria bassiana infect and kill pests such as aphids and whiteflies. Nematodes: Parasitic nematodes attack pest larvae in the soil, such as root grubs. (8) **Plant-Insect Interactions**: Trap Cropping: Certain plants attract pests away from main crops, reducing damage. Examples: Planting mustard as a trap crop for diamondback moths in cabbage fields. Push-Pull System: Combining repellent crops ("push") and attractive trap crops ("pull") to manage pests. Example: Desmodium repels stem borers while Napier grass attracts them in maize farming. (9) **Herbivory**: Certain herbivores feed on pest weeds that harbor harmful insects, reducing pest habitats. Examples: Goats: Feed on invasive weeds, reducing pest habitat.

Natural Control of Diseases: Interspecific interactions play a significant role in the natural control of plant diseases in the natural farming agroecosystem. These interactions occur between different species and help suppress plant pathogens, enhance plant health, and maintain ecological balance. The key types of interspecific interactions involved in natural disease control include: (1) **Predation**: Predators feed on disease-causing organisms, reducing their population. Examples: Nematode-Trapping Fungi: Predatory fungi like Arthrobotrys capture and consume nematodes that harm plant roots. Lady Beetles: Prey on aphids and other pests that act as vectors for plant pathogens. Spiders: Feed on insects that may carry fungal or bacterial pathogens. (2] **Competition**: Beneficial microbes compete with pathogens for resources such as nutrients, space, or ecological niches, thereby suppressing disease development. Examples: *Trichoderma* spp. competes with pathogenic fungi like Fusarium or Rhizoctonia, preventing their growth. *Pseudomonas fluorescens* competes with soil-borne pathogens by producing siderophores that sequester iron, limiting pathogen survival. (3) **Parasitism**: Parasites live on or in disease-causing organisms, weakening or killing them. Examples: *Trichoderma harzianum*: Parasitizes fungal pathogens by penetrating their hyphae and disrupting their structure. *Ampelomyces quisqualis*: A hyperparasite that infects powdery mildew fungi. (4) Mutualism: Both species benefit from the interaction, often leading to enhanced plant health and reduced disease susceptibility. Examples: Mycorrhizal Fungi and Plants: Mycorrhizal fungi enhance nutrient uptake for plants and, in return, receive carbohydrates. They also create a physical and chemical barrier against soil pathogens. Nitrogen-Fixing Bacteria (e.g., Rhizobia): Improve plant health, making them less susceptible to diseases. (5) **Commensalism**: One species benefits while the other is unaffected, but the interaction indirectly aids in disease control. Examples: Epiphytic Bacteria: Non-pathogenic bacteria on plant surfaces inhibit the colonization of harmful pathogens. 6) **Amensalism**: One species is harmed while the other is unaffected, often through the production of harmful substances. Examples: Antibiotic production by *Streptomyces* bacteria inhibits the growth of fungal pathogens. Fungi like *Aspergillus* produce compounds that suppress the growth of other pathogens. (7) **Plant-Insect Interactions**: Pollinators and Beneficial Insects: Pollinators like bees indirectly reduce diseases by ensuring healthy crop growth. Insect Predators and Parasitoids: Wasps parasitize pest larvae that vector diseases, breaking the pathogen-pest cycle. (8) **Allelopathy**: Some plants release chemicals (allelochemicals) that inhibit the growth of pathogens. Examples: Marigolds release compounds that suppress nematodes and soil-borne pathogens. Mustard plants produce glucosinolates that reduce pathogen populations in the soil. (9) **Induced Systemic Resistance (ISR):** Beneficial microbes or insects induce plant defenses, making them resistant to pathogens. Examples: *Pseudomonas fluorescens* induces systemic resistance in plants, enhancing their ability to combat pathogens. *Bacillus subtilis* stimulates the production of plant defense enzymes like chitinases and peroxidases.

Natural Control of Weeds: In natural farming, interspecific interactions play a vital role in controlling weeds. These interactions involve relationships between crops, weeds, microorganisms, insects, and other organisms that suppress weed growth or limit their impact on crops. Key types of interspecific interactions involved in natural weed control include: (1) **Competition:** Crops compete with weeds for resources such as sunlight, water, nutrients, and space, reducing weed growth. Examples: Dense planting or fast-growing crops like maize or barley outcompete weeds by shading them. Intercropping legumes with cereals suppresses weeds by creating less space and fewer resources for their growth. (2) **Allelopathy:** Certain plants release biochemical compounds (allelochemicals) into the soil that inhibit weed germination and growth. Examples: Sorghum: Releases sorgoleone, which suppresses broadleaf and grassy weeds. Mustard and Brassicas: Produce glucosinolates that suppress weed seed germination. Rice Straw: Residue left after harvest can inhibit weed growth through allelopathic compounds. (3) **Predation:** Natural predators consume weed seeds or seedlings, reducing weed populations. Examples: Ground Beetles: Feed on weed seeds, especially of broadleaf weeds. Birds: Forage on weed seeds in fields, reducing their spread. (4) **Parasitism:** Certain parasitic organisms attach to or infect weeds, reducing their growth and reproduction. Examples: Parasitic Fungi: Fungi like Puccinia spp. infect and weaken weeds, such as wild grasses. Cuscuta (Dodder): Although a parasitic plant itself, it can suppress specific aggressive weeds under controlled use. (5) **Mutualism:** Beneficial interactions between crops and other organisms indirectly control weeds. Examples: Mycorrhizal Fungi: Enhance nutrient uptake in crops, making them more competitive against weeds. Nitrogen-Fixing Bacteria: Strengthen legumes in intercropping systems, outcompeting nitrogen-dependent weeds. (6) **Commensalism:** One species benefits while the other remains unaffected, indirectly helping in weed control. Examples: Epiphytic plants or microorganisms that inhabit crop surfaces without harming them may outcompete or inhibit weed colonization in the vicinity. (7) Herbivory: Grazing animals feed on weeds, controlling their populations. Examples: Livestock like goats or sheep can graze on weeds in pastures. Aquatic weeds like water hyacinth are consumed by fish species such as grass carp. (8) **Cover Cropping and Mulching:** Cover crops and mulch physically suppress weed growth and interact biologically with the soil and crop ecosystem. Examples: Legume Cover Crops: Compete with weeds and improve soil fertility. Dead Mulch: Organic residues create a physical barrier and decompose, releasing compounds that suppress weed growth. (9) **Integrated Crop-Weed Interaction Management:** Intercropping: Planting multiple crops together creates competition and reduces space for weeds to thrive. Crop Residue Management: Retaining residues on fields suppresses weed emergence while nourishing the soil. Companion Planting: Some plants naturally repel or suppress weeds when grown together. (10) **Biological Control Agents:** Using organisms that naturally suppress weeds through predation, parasitism, or competition. Examples: Insects: *Zygogramma bicolorata* feeds on the leaves of the invasive weed *Parthenium hysterophorus*. *Neochetina eichhorniae* and *Neochetina bruchi* control water hyacinth (*Eichhornia crassipes*). Pathogens: *Alternaria eichhorniae* and *Fusarium* spp. infect and suppress specific weeds. (11) **Role of Microbial Interactions:** Soil microbes influence weed suppression by competing for nutrients or producing toxic metabolites. Examples: *Pseudomonas fluorescens*:

Suppresses weed seedling growth by competing for iron through siderophore production. Rhizobacteria: Some strains inhibit weed germination and growth by releasing phytotoxins.

9.1.2 Conservation of Natural Capital in Natural Farming Agroecosystem:

Natural capital refers to the stock of natural resources—such as soil, water, biodiversity, and air—that supports life and agriculture. In natural farming, conserving these resources ensures long-term sustainability, minimizes external inputs, and enhances resilience to environmental changes. Key Components of Natural Capital in natural farming agroecosystems: (1) **Soil Health:** Organic matter, soil microbes, and nutrient cycles that support plant growth. (2) **Water Resources:** Groundwater, surface water, and rainfall that provide irrigation and maintain soil moisture. (3) **Biodiversity:** Diverse plant species, beneficial insects, pollinators, and soil organisms. (4) **Air and Climate Regulation:** Atmospheric processes that support photosynthesis and moderate temperature. (5) **Energy Resources:** Sunlight and other renewable energy sources critical for ecosystem functioning.

9.1.3 Phytoremediation of Environmental Pollution:

Phytoremediation is an eco-friendly technology that utilizes plants to clean contaminated soil, water, and air. It encompasses various processes, including: **Phytoextraction:** Plants absorb and accumulate pollutants in their tissues. **Phytodegradation:** Plants break down organic contaminants into less harmful substances. **Phytostabilization:** This process immobilizes contaminants in the soil to prevent their spread. **Rhizofiltration:** Roots filter out pollutants from water.

9.1.4 *Maximising Human Health and Immunity:*

The health of plants directly influences human health and immunity, as plants are primary sources of essential nutrients, bioactive compounds, and environmental benefits that contribute to overall well-being. (1) **Nutritional Quality of Food:** Healthy plants produce nutrient-rich fruits, vegetables, and grains that are vital for human health. Key nutrients include: Vitamins (A, C, E): Boost immunity and prevent infections. Minerals (Iron, Zinc, Selenium): Support immune functions and reduce deficiencies. Antioxidants (Flavonoids, Carotenoids): Protect cells from damage and strengthen the immune system. Example: Dark green leafy vegetables, when grown in nutrient-rich soil, provide higher levels of iron and folate essential for immunity. (2) **Bioactive Compounds:** Plants produce secondary metabolites like: Polyphenols: Found in berries, tea, and cocoa, improve gut health and immune response. Alkaloids: Strengthen the body's defense mechanisms against diseases. Terpenes: Present in herbs like basil and thyme, offer antimicrobial and anti-inflammatory properties. These compounds help in managing chronic diseases and boosting immunity. (3) **Reduced Exposure to Harmful Chemicals:** Plants grown without synthetic pesticides and fertilizers are safer for consumption. Reduced chemical residues in food lower the risk of chronic diseases, hormonal imbalances, and immune suppression. (4) **Enhanced Gut Health:** Dietary fiber from plants supports the growth of beneficial gut bacteria. A healthy gut microbiome is crucial for a robust immune system as it regulates inflammation and pathogen defense. (5) **Environmental Benefits that Impact Human Health:** Healthy plants in natural ecosystems improve air and water quality. Forests and green spaces reduce pollution and promote mental health, indirectly boosting immunity. (6) **Medicinal Plants and Natural Remedies:** Many plants with medicinal properties enhance immunity and combat diseases. Examples: Turmeric: Contains curcumin, which has strong anti-inflammatory and immune-boosting effects. Ginger: Known for its antiviral and antibacterial properties. Garlic: Rich in sulfur compounds that enhance immune cell function. (7) **Reduced Stress and Improved Mental Health**: Green environments and access to fresh, healthy plants reduce stress, which is a known suppressor of the immune system. Activities like gardening and consuming fresh produce promote mental and physical well-being. (8) **Mitigation of Zoonotic Diseases**: Healthy agroecosystems reduce the likelihood of disease transmission from animals to humans by maintaining ecological balance. Example: Crop diversification and pest management through natural means minimize the use of antibiotics and prevent antibiotic resistance.

9.1.5 Minimising Risk and Cost of Crop Production:

The objective of plant health management in natural farming is to minimise the risk and cost of production of crops. Plant health management through improvement of soil health and improvement of biodiversity automatically reduces risk by diversification and reduces cost by avoiding hybrid seeds, fertilisers and pesticides.

9.1.6 Maximising Aggregate Yield of Crops:

The objective of plant health management in natural farming agroecosystem is to maximise the aggregate yield of crops per year. This is achieved through rating and ranking of plant species and varieties, optimising the crop plant portfolio and designing the farm landscape (multilayer polycropping system).

9.1.7 Maximising Quality of the Agricultural products:

The objective of plant health management in natural farming is to maximise the quality of agricultural products. Quality of crops can be improved through plant breeding, agroecosystem management and pollination management.

9.1.8 Maximising Profit of the Farmers:

The ultimate objective of plant health management in natural farming is to maximise the profit of the farmers.

9.2 Methods of Plant Health Management:

Methods of plant health management in natural farming are completely different from conventional agriculture. It involves genetic improvement of plant varieties, soil health improvement, improvement of plant defense against biotic and abiotic stresses and natural control of insect pests, diseases and weeds.

Plant Breeding for Natural Farming: Farmers traditionally do plant domestication and plant breeding for improvement of their adaptability to their natural environmental conditions, improvement of growth and reproduction under the conditions of natural farming (no-till, no-inputs, no-irrigation), and improvement of their tolerance, resistance and immunity against insect pests, diseases and weeds (See Chapter 4).

Farm Landscape Design: Farmers design their multilayer polycropping systems wisely to maintain proper spacing (plant to plant distance), stacking the plant canopies in 5 to 7 layers according to plant heights, minimising shading effects on lower level crops, proper mixing of crops to increase plant nutrition and to reduce incidence of pests, diseases and weeds, to improve plant health. The ultimate objective is to create biodiversity at the genetic level (plant varieties and animal breeds), at species level (crops and livestocks) and at ecosystem level (cropland, grassland, orchard, agroforestry, pond, hedgerow, biofencing).

Crop Succession: Annual crops have specific sowing or planting time for achieving optimum plant growth and reproduction. Traditional farmers sow the seeds of different crops in succession within the standing crops (relay cropping, companion cropping or poira cropping) to maximise the cropping Intensity and to maximise the utilisation of available resources to improve plant health.

Water Cycle Management: See chapter 7.2.

Soil health management: See chapter 8.2.

Probiotics:

Traditional home-made crude microbial cultures (probiotics) are natural preparations made using locally available materials and indigenous knowledge to promote plant health, enhance soil fertility, and manage pests and diseases sustainably. These microbial cultures harbor beneficial microorganisms like bacteria, fungi, and yeasts that support plant growth, suppress pathogens, and improve resilience against environmental stresses.

Benefits of Traditional Crude Microbial Cultures: (1) **Soil Fertility:** Decompose organic matter to release essential nutrients. Improve soil structure and microbial diversity. (2) **Plant Growth:** Stimulate root development and enhance nutrient uptake. Produce phytohormones like auxins and gibberellins. (3) **Pest and Disease Management:** Compete with harmful pathogens, reducing their population. Produce antimicrobial compounds that suppress diseases. (4) **Eco-friendly and Cost-effective:** Use locally available materials, reducing costs for farmers. Eliminate the need for synthetic fertilizers and pesticides.

Popular Home–Made Crude Microbial Cultures: (1) **Jeevamrut** (India): Ingredients: Cow dung, cow urine, jaggery, gram flour, and water. Preparation: Mix ingredients in water and ferment for 5–7 days. Usage: Soil application or foliar spray; promotes nutrient cycling and suppresses soil-borne diseases. (2) **Panchagavya** (India): Ingredients: Cow dung, cow urine, milk, curd, ghee, banana, jaggery, and tender coconut water. Preparation: Ferment the mixture for 7–10 days. Usage: Enhances plant immunity, growth, and yield. (3) **Fermented Plant Juice** (FPJ): Ingredients:

Fresh plant leaves (e.g., bamboo, banana), brown sugar or jaggery, and water. Preparation: Layer leaves with jaggery in a container and ferment for 5–7 days. Usage: Provides nutrients and plant growth-promoting microorganisms. (4) **Fish Amino Acid** (FAA): Ingredients: Fish waste, jaggery or sugar, and water. Preparation: Mix ingredients and ferment for 1–2 weeks. Usage: Rich in nitrogen and promotes vegetative growth. (5) **Effective Microorganisms** (EM): Ingredients: Lactic acid bacteria, yeast, and photosynthetic bacteria with molasses or jaggery. Preparation: Ferment in an anaerobic environment for 7–10 days. Usage: Improves soil health and suppresses pathogens. (6) **Compost Tea:** Ingredients: Well-decomposed compost, water, and molasses or jaggery. Preparation: Aerate the mixture for 24–48 hours. Usage: Enhances soil microbial activity and plant nutrient availability. (7) **Ginger-Garlic Extract:** Ingredients: Crushed ginger, garlic, jaggery, and water. Preparation: Ferment for 3–5 days. Usage: Acts as a natural pesticide and fungal suppressant.

Plant Geometry Management: Farmers follow different methods of grafting, pruning and training to optimise the plant health and the plant geometry, to maximise yield and quality of crops.

Weed Management: Farmers design their multilayer polycropping system wisely to minimise the occurrence of weeds in their natural farming agroecosystems, through strategic use of shading effect and allelopathy. This passive method of weed control is free-of-cost and help profit maximisation of the farmers.

Insect Pest Management: Farmers wisely create plant biodiversity in the multilayer polycropping system of the natural farming agroecosystems, to manage the population dynamics of the pests and their natural enemies (predators and parasitoids), automatically and permanently, below the economic threshold level. This system therefore help improve plant health and help profit maximisation of the farmers.

Disease Management: Farmers breed disease resistant plant varieties and manage plant biodiversity in the natural farming agroecosystem to maximise plant health and to manage plant diseases with the use of antagonists, probiotics, epiphytes. This system not only maintains plant health free-of-cost but helps maximise profit of the farmers.

9.3 Interspecific Interactions in Plant Health Management:

9.3.1 Arbuscular Mycorrhizal Fungi (AMF):

Arbuscular Mycorrhizal Fungi (AMF) play a vital role in enhancing plant health and productivity through a symbiotic relationship with plant roots. They improve nutrient uptake, enhance stress tolerance, and contribute to sustainable agricultural practices. Here's a detailed look at their role in plant health:

Enhanced Nutrient Uptake:

Phosphorus Absorption: AMF increase the surface area of plant roots through hyphal networks, enhancing the uptake of immobile nutrients like phosphorus. **Other Nutrients:** They also aid in the uptake of nitrogen, zinc, copper, and potassium.

Improved Soil Structure:

Soil Aggregation: AMF produce glomalin, a glycoprotein that binds soil particles, improving soil structure and reducing erosion. **Enhanced Water Retention:** Better soil aggregation increases water-holding capacity, supporting plant growth in dry conditions.

Stress Tolerance:

Drought Resistance: AMF help plants withstand water stress by improving water uptake efficiency. **Salt Tolerance:** They reduce the harmful effects of salinity by regulating ion balance in plant roots. **Heavy Metal Tolerance**: AMF immobilize toxic heavy metals in the soil, protecting plants from their harmful effects.

Disease Resistance:

AMF colonization reduces the incidence of soil-borne diseases by: Competing with pathogenic fungi and bacteria for space and resources. Activating the plant's systemic resistance mechanisms.

Carbon and Energy Exchange:

Photosynthesis Efficiency: AMF improve nutrient supply, leading to better photosynthetic performance and energy production in plants. **Carbon Storage in Soil**: AMF contribute to carbon sequestration by stabilizing organic matter in the soil.

Plant Growth and Development:

AMF enhance root growth and development, leading to better anchorage and increased biomass. They also influence shoot growth by improving nutrient and water availability.

Biodiversity and Ecosystem Services:

Microbial Diversity: AMF support diverse microbial communities in the rhizosphere, promoting a balanced soil ecosystem. **Plant Biodiversity**: They enable coexistence of plant species by facilitating nutrient sharing in mixed cropping systems.

9.3.2 Plant Growth Promoting Rhizobacteria (PGPR):

Plant Growth-Promoting Rhizobacteria (PGPR) are beneficial soil bacteria that colonize plant roots and enhance plant health by improving growth, nutrient availability, and resistance to biotic and abiotic stresses. They are essential for sustainable agriculture and natural farming. Here's a detailed explanation of their role in plant health:

Nutrient Mobilization and Uptake: Nitrogen Fixation: Symbiotic PGPR (e.g., Rhizobium species) fix atmospheric nitrogen into forms usable by plants. Free-living nitrogen fixers (e.g., Azotobacter, Azospirillum) contribute to nitrogen availability. **Phosphorus Solubilization:** PGPR solubilizes insoluble phosphates into forms plants can absorb (e.g., Bacillus, Pseudomonas). **Potassium and Zinc Mobilization:** PGPR release enzymes and acids that make these nutrients more available to plants.

Production of Plant Hormones: Auxins (e.g., IAA): Stimulate root elongation and branching, enhancing nutrient and water absorption. **Cytokinins:** Promote cell division and shoot growth. **Gibberellins:** Aid in seed germination, stem elongation, and flowering.

Disease Suppression: Antibiotic Production: PGPR produce antibiotics that inhibit pathogenic bacteria and fungi (e.g., *Pseudomonas fluorescens*). **Induction of Systemic Resistance (ISR):** Trigger plant defense mechanisms, enhancing resistance to pathogens. **Competition for Nutrients and Space:** Outcompete harmful microbes in the rhizosphere.

Abiotic Stress Tolerance: Drought Resistance: PGPR produce exopolysaccharides and increase root hydraulic conductivity, improving water retention and uptake. **Salt Tolerance:** Help maintain ion balance and produce osmoprotectants, reducing the impact of salinity. **Heavy Metal Detoxification:** PGPR sequester or transform toxic metals, reducing their availability to plants.

Enhanced Soil Health: Rhizosphere Dynamics: Improve microbial diversity and activity in the root zone. **Organic Matter Decomposition:** Break down organic residues, enriching soil fertility.

Biocontrol of Pests: PGPR like *Bacillus thuringiensis* produce toxins that target insect pests. Indirectly reduce pest damage by promoting healthier, more resilient plants.

Enhanced Seed Germination and Early Growth: Seed coating with PGPR promotes faster germination and stronger early root and shoot growth.

Carbon Sequestration and Environmental Benefits: By enhancing root biomass, PGPR contributes to increased carbon sequestration in the soil.

9.4 Biomonitoring of plant health:

Plant health monitoring involves the systematic observation and analysis of plant conditions to ensure they are growing healthily and to detect early signs of stress or disease. Techniques for Plant Health Monitoring: **Visual Inspection**: Regularly checking plants for discoloration, spots, wilting, or other visible symptoms of stress or disease. **Soil Analysis**: Monitoring soil pH, moisture, and nutrient levels to ensure optimal growing conditions. **Leaf and Stem Analysis**: Observing leaf color, texture, and stem integrity for signs of nutrient deficiencies or pest infestations. **Remote Sensing**: Using drones, satellites, or cameras equipped with multispectral or hyperspectral imaging to detect issues like chlorophyll levels or water stress. **IoT Sensors**: Employing devices that measure soil moisture, temperature, humidity, and nutrient levels in real-time. **Plant Sap Analysis**: Testing the plant's internal fluids to assess its nutrient uptake and detect stressors.

CHAPTER 10

POLLINATION MANAGEMENT IN NATURAL FARMING AGROECOSYSTEM

Pollination: Pollination is the transport of male gamete from the anther of a flower to the pistil of a female gamete of a flower through air, water or biotic means to cause successful fertilisation. Farmers try to maximise biotic pollination of crop plants efficiently to maximise crop yield, crop quality and price of the agricultural products. Farmers maximise pollinator biodiversity in the natural farming agroecosystems, by collecting maximum numbers of pollinator magnet crops and making the natural farming agroecosystem a pollinator sanctuary or pollinator hotspot. Farmers also practice beekeeping (apiculture) to maximise biotic pollination of crops as well as for producing honey, a valuable agricultural product.

10.1 Objectives of Pollinator Management:

Farmers should actively manage their native pollinator biodiversity in their natural farming agroecosystems, to effectively improve crop production and crop protection. The main objectives of pollinator management include:

10.1.1 To Maximise Pollinator Biodiversity:

The first objective of pollinator management, is to maximise the pollinator biodiversity in the natural farming agroecosystem. Pollinator biodiversity in agriculture refers to the variety of pollinating species that contribute to crop production and ecosystem health. Examples include: (1) **Insects: (a) Bees: Honeybees:** There are several species of honeybees worldwide, each with distinct characteristics and ecological roles. **Apis mellifera** (Western Honeybee). This is the most common and widely domesticated species of honeybee. It is found across Europe, Asia, Africa, and has been introduced worldwide. It is highly productive in terms of honey production and pollination, making it the most commercially important species. There are several subspecies of *Apis mellifera*, such as *Apis mellifera* carnica (Carniolan honeybee), *Apis mellifera ligustica* (Italian honeybee), and *Apis mellifera scutellata* (Africanized honeybee). (2) **Apis cerana** (Asian Honeybee): Native to South and Southeast Asia, *Apis cerana* is a smaller bee compared to *Apis mellifera*. It has adapted to a variety of climates and conditions, including tropical environments.

It plays a vital role in pollination of many crops, including fruit trees and oilseed crops. It produces less honey than *Apis mellifera*, but the honey it produces is valued for its medicinal properties in some cultures. (3) **Apis dorsata** (Giant Honeybee): This species is known for its large size and ability to build open, exposed nests in tall trees or cliffs. Found across Southeast Asia, it is a wild honeybee species. It pollinates large flowering plants, particularly in tropical rainforests. (4) **Apis florea** (Dwarf Honeybee): *Apis florea* is a smaller species of honeybee found in Southeast Asia. It nests in low shrubs or trees, and its colonies are relatively small. It is primarily a wild species and not commonly used for commercial honey production. Plays a significant role in pollinating tropical crops and flowering plants, particularly in regions where other species may not thrive. (5) **Apis mellifera saharae** (Saharan Honeybee): A subspecies of *Apis mellifera*, this honeybee is adapted to the harsh desert environments of North Africa, particularly the Sahara region. (6) **Apis nigrocincta** (Philippine Honeybee): Found mainly in the Philippines and some nearby Southeast Asian islands, this species is similar to *Apis cerana* but has distinct behaviors and hive structures. (7) **Apis koschevnikovi** (Indochinese Honeybee): A species native to Southeast Asia, particularly the Indo-Chinese region. It is smaller than *Apis mellifera* and typically nests in trees or caves. (8) **Apis**

cerana indica (Indian Honeybee): A subspecies of *Apis cerana* found in India. It is adapted to the sub-tropical and tropical climates of the Indian subcontinent. Pollinates a wide range of crops, especially fruits, vegetables, and oilseeds. (9) **Apis andreniformis** (Black Dwarf Honeybee): A small, wild species found in Southeast Asia. It is less known compared to other species but contributes to the pollination of tropical plants. (10) **Apis breviligula** (Japanese Honeybee): Native to Japan, this species is smaller than the Western honeybee and adapted to the Japanese climate. **Bumblebees** (*Bombus* spp.): Excellent for pollinating crops like tomatoes and peppers, as they perform buzz pollination. **Stingless Bees** (Meliponini): Important in tropical regions for crops like coffee and mango. (b) **Butterflies**: Pollinate flowers of crops like carrots, radishes, and some fruits. Examples include monarchs and swallowtails. (c) **Moths**: Nocturnal pollinators for crops like evening primrose and certain fruits. (d) **Hoverflies**: Contribute to pollination of crops like lettuce, onions, and brassicas. (e) **Beetles**: Pollinate ancient crops like custard apple and magnolia. (f) **Wasps**: Though less efficient, they assist in pollination of figs and some flowers. (2) **Birds**: (a) **Hummingbirds**: Pollinate flowers of crops like passionfruit and guava in the Americas. (b) **Sunbirds**: Active in pollination in tropical areas, such as in coconut and papaya. (3) **Mammals: Bats:** Essential for pollinating night-blooming crops like bananas, guavas, and agaves (for tequila). (4) Other Pollinators: **Ants**: Pollinate low-growing crops and flowers like wild strawberries. **Beetles**: Known as "mess and soil" pollinators, they assist crops with sturdy flowers like palms. (5) **Aquatic Pollinators: Water Beetles:** Pollinate some aquatic plants like water lilies, indirectly supporting wetland agriculture systems.

10.1.2 To Maximise Pollination Efficiency:

The pollination efficiency of a pollinator is influenced by various biological, ecological, and environmental factors. (1) **Foraging**

Strategy of Pollinators: (a) **Foraging Behavior:** Pollinators like *Apis dorsata, Apis mellifera,* and *Apis cerana* show distinct foraging behaviors that influence their pollination efficiency. (b) **Efficient Foragers:** *Apis dorsata* and *Apis mellifera* exhibit a higher relative pollination efficiency (RPE) due to their consistent foraging routes and effective pollen transfer. (c) **Pollen Theft:** Some pollinators may engage in pollen theft (removing pollen without effective pollination), which can reduce the efficiency of the overall pollination process. (d) **Pollen Load:** The quantity of pollen carried by pollinators is crucial. Larger bees often carry more pollen, improving pollination efficiency. (2) **Flowering Phenology:** (a) **Blooming Time and Synchronization:** The timing of flower blooming and its synchronization with the activity period of pollinators is essential. For example: (b) **Morning Activity:** Pollinators like honeybees are most active during morning hours, coinciding with peak anthesis (pollen release). (c) **Staggered Blooming:** Plants with staggered flowering periods ensure prolonged pollination opportunities. (d) **Anthesis Duration:** The duration of flower receptivity (when stigmas are fertile) impacts pollinator visits and, consequently, fertilization rates. (3) **Flower Characteristics:** (a) **Shape:** Tubular flowers are better suited for long-tongued pollinators like butterflies. Open, flat flowers attract generalist pollinators like bees. (b) **Nectar and Pollen Content:** Flowers with abundant nectar and pollen are more attractive to pollinators. Nectar sugar concentration can influence the choice of pollinators (e.g., honeybees prefer sucrose-rich nectar). (c) **Color:** Brightly colored flowers (e.g., yellow, blue, and purple) attract specific pollinators like bees, while red flowers attract birds. (4) **Flower Arrangement:** (a) **Aggregation:** Clusters of flowers are more efficient in attracting pollinators as they reduce the energy spent by pollinators in foraging. Examples include crops like sunflower and apple ber, where flower aggregation increases pollination efficiency. (5) **Biodiversity of Pollinator Species:** (a) **Species Diversity:** A higher diversity of pollinator species ensures better pollination, as different pollinators target flowers at varying times or in different ways. (b) **Functional Complementarity:** Some pollinators are specialized for nectar collection, while others focus on pollen collection, ensuring diverse pollination services. (6) **Interspecific Interactions:** (a) **Coevolution Theory:** Pollinators and plants often coevolve, developing traits that optimize pollination efficiency. Example: Plants with deep corollas may evolve alongside pollinators with long proboscises. (b) **Competition:** Competition among pollinator species can influence foraging behavior. Aggressive species like *Apis dorsata* may dominate resources, impacting less aggressive species. (7) **Environmental Factors:** (a) **Daytime:** Pollinator activity is typically highest during mid-morning to early afternoon due to optimal temperature and light conditions. (b) **Wind Velocity:** High wind velocity can deter pollinator activity, reducing efficiency. In contrast, calm conditions are conducive to effective pollination. (c) **Temperature and Humidity:** Optimal temperature and humidity levels ensure higher pollinator activity. For instance, bees are most active at moderate temperatures. (8) **Distance from Hive to Flower:** (a) **Foraging Range:** The distance between the hive and the flowers impacts energy expenditure and foraging efficiency. Bees like *Apis dorsata* can forage over longer distances (up to 5 km) compared to *Apis cerana* (1-2 km). (b) **Closer Proximity:** Shorter distances increase the frequency of pollinator visits, enhancing pollination efficiency. For example maximum pollination efficiency in mango with *Apis cerana* is 5m. (9) **Direction of the Flower:** (a) **Accessibility:** Flower orientation affects the ease with which pollinators can access nectar and pollen. Downward-facing flowers may be more challenging for some pollinators, while upward-facing flowers are more accessible. (10) **Predators and Parasites:** (a) **Impact on Behavior:** The presence of predators like spiders or parasites in flowers can deter pollinators, reducing their activity. Example: Hoverflies may avoid flowers where predatory spiders are present. (b) **Defensive**

Pollinators: Some species like *Apis dorsata* are more resilient to predators, ensuring consistent pollination. (11) **Competition Among Plants**: (a) **Resource Partitioning**: When multiple plant species bloom simultaneously, competition for pollinator visits can affect individual plant pollination rates. Plants with more attractive traits (e.g., brighter flowers or higher nectar content) may dominate, reducing pollination for less attractive species.

10.1.3 To Maximise crop yield:

Biotic pollination plays a crucial role in increasing crop yield. (1) **Improved Fertilization**: Efficient pollen transfer by pollinators increases the chances of fertilization. Ensures the development of more flowers into fruits or seeds. (2) **Higher Fruit and Seed Set**: Pollinators ensure better pollen deposition, leading to a higher proportion of flowers setting fruit or seeds. Cross-pollination enhances genetic diversity, which often improves fruit set. (3) **Enhanced Fruit and Seed Quality**: Biotic pollination results in larger, more uniform fruits and seeds. Improves the nutritional content and marketability of the produce. (4) **Increased Pollination Efficiency**: Biotic agents can access hard-to-reach parts of flowers, ensuring effective pollination. For self-incompatible species, pollinators are essential for cross-pollination, as these plants cannot self-fertilize. (5) **Prolonged Flower Viability**: Frequent visits by pollinators stimulate flowers to remain open and viable for longer, increasing the chances of successful pollination. (6) **Enhanced Crop Yield Stability**: Consistent pollinator activity reduces the variability in yield, ensuring more reliable production across seasons. (7) **Boosting Yield in Mixed Cropping** (Polyculture): In polyculture systems, biotic pollination facilitates effective pollination across multiple crops simultaneously, maximizing overall yield. (8) **Reduction in Crop Loss**: Pollinators can overcome environmental constraints like wind or low insect activity during specific periods, ensuring reliable pollination. (9] **Secondary Effects**: Pollinators like bees can contribute to improved pest control by feeding on pests or attracting pest predators, indirectly improving crop yield. Examples: (1) **Apple Orchards:** Managed honeybees (*Apis mellifera*) increased fruit set and size, improving yield by up to 30%. (2) **Sunflower Fields:** Native pollinators like bumblebees enhance the seed set, leading to 20–25% higher yields compared to wind pollination alone. (3] **Tomato Cultivation:** Vibratory pollination by bumblebees resulted in larger and more uniform fruits compared to artificial methods.

10.1.4 To Maximise Quality of Crops:

Biotic pollination, carried out by living organisms such as bees, butterflies, birds, and bats, not only increases crop yield but also significantly enhances crop quality. The quality improvements are observed in terms of fruit size, shape, taste, nutritional content, and market value. Biotic pollinators ensure effective pollen transfer, which leads to better fertilization and subsequent development of high-quality produce. (1) **Uniform Fruit and Seed Development:** Proper pollination ensures even fertilization of ovules, resulting in uniformly developed fruits and seeds. Poor or incomplete pollination often causes misshapen or deformed fruits. Example: Apples pollinated by bees are rounder and more evenly shaped, making them more marketable. (2) **Larger Fruit and Seed Size:** Cross-pollination by biotic agents increases the number of fertilized ovules, leading to larger fruits and seeds. Larger size often translates to higher commercial value. Example: Watermelons pollinated by bees tend to be larger compared to those relying on self-pollination or wind. (3) **Improved Taste and Flavor:** Enhanced fertilization results in higher sugar accumulation in fruits, improving sweetness and overall taste. Certain crops exhibit better aroma and flavor profiles due to effective biotic pollination. Example: Strawberries pollinated by bees are sweeter and have a better flavor than poorly pollinated ones. (4] **Higher Nutritional Content:** Fruits and seeds produced through biotic pollination often have higher levels of essential nutrients such as vitamins, proteins, and oils. The enhanced nutrient profile is linked to better pollen transfer and fertilization. Example: Sunflower seeds pollinated by bees have a higher oil content compared to seeds from wind-pollinated plants. (5) **Longer Shelf Life**: Properly pollinated fruits are less prone to premature rotting or mechanical damage, resulting in better storage and transportability. Uniformity in ripening also helps extend shelf life. Example: Biotically pollinated tomatoes show delayed decay compared to hand-pollinated ones. (6) **Reduced Crop Defects**: Biotic pollination reduces the incidence of hollow seeds or incomplete fruit formation, commonly caused by inadequate fertilization. The absence of defects enhances the cosmetic appeal of the product. Example: Cucumbers pollinated by bees have fewer hollow centers and a smoother surface. (7) **Improved Seed Viability**: Seeds resulting from biotic pollination are more viable and exhibit better germination rates, contributing to better propagation and future crop success. Viable seeds are essential for the production of high-quality crops in subsequent generations. Example: Canola seeds pollinated by bees have higher germination percentages compared to self-pollinated seeds. (8) **Enhanced Crop Diversity**: Biotic pollinators encourage cross-pollination, increasing genetic diversity within crops. This diversity can lead to better resilience against diseases and environmental stresses. Example: Cross-pollinated fruits like papayas have higher resistance to certain pests and diseases.

10.1.5 To Repel Wild Animals:

Honey bees can repel wild animals through their natural defensive behavior and the unique chemical and physical characteristics of their hive. This trait makes them valuable for deterring crop-damaging animals and protecting agricultural fields. The following mechanisms explain how honey bees repel wild animals: (1)

Aggressive Defense Behavior: Stinging as a Deterrent: Honey bees aggressively defend their hive or surrounding area when threatened. Their stings, though painful, release alarm pheromones that attract other bees to attack, effectively repelling larger animals like elephants, bears, and wild boars. Example: Elephants avoid areas with active bee colonies because even a few bee stings in sensitive areas (e.g., inside the trunk) can cause significant discomfort. **Chasing Intruders:** Bees often pursue animals that disturb their hive or forage near it, creating a strong psychological deterrent. (2) **Alarm Pheromones:** Honey bees release alarm pheromones (like isopentyl acetate) when they sting or feel threatened. These pheromones signal other bees to swarm the intruder, making the environment intimidating for wild animals. Animals that have experienced bee attacks before tend to associate the scent with danger and avoid such areas. (3] **Sensory Discomfort:** The buzzing sound of bees is often enough to scare wild animals. The sound signals potential danger to animals with sensitive hearing, such as elephants or deer. This fear response is an evolutionary adaptation, as animals have learned to associate bee activity with painful stings. (4) **Hive as a Natural Barrier:** Honey bee colonies placed around fields act as a "living fence" to deter animals like wild boars, monkeys, or even birds. Bees patrol the area near their hives and instinctively attack when they perceive movement or vibrations. (5) **Repelling Elephants** (Special Case): Elephants are particularly sensitive to bees and actively avoid areas with buzzing sounds or pheromones. Farmers in regions like Africa and India use "beehive fences" to protect crops. These fences consist of connected beehives that swing and disturb the bees when an elephant approaches, causing the bees to swarm and scare the animal away. (6) **Honey Bees in Agriculture and Wildlife Management: Beehive Fences:** A sustainable method used by farmers in elephant-prone areas. The beehives are spaced at intervals along the field's boundary and connected by wires. When a large animal touches the wire, it shakes the hives, activating the bees. **Intercropping with Bee-friendly Plants**: Planting nectar-rich crops or wildflowers around farms attracts bees and creates a natural barrier.

10.2 Strategies to Maximise Pollinator Biodiversity in Natural Farming Agroecosystem:

Maximizing pollinator biodiversity in a natural farming agroecosystem is crucial for improving crop yields, maintaining ecological balance, and supporting long-term farm sustainability. Here are the best strategies to attract and sustain a diverse range of pollinators:

10.2.1 Plant Diverse Flowering Crops:

Grow a variety of flowering crops that bloom at different times of the year to provide continuous nectar and pollen sources. Supports a wide range of pollinator species throughout the year. Include sunflowers, clovers, mustard, and legumes in crop rotation or as cover crops.

10.2.2 Create Pollinator-Friendly Habitats:

Establish natural habitats like hedgerows, wildflower strips, or grasslands near farmlands. Provides nesting sites and food for pollinators. Leave patches of bare ground for ground-nesting bees. Provide deadwood for cavity-nesting bees.

10.2.3 Avoid Pesticide Use:

Eliminate synthetic pesticide use. Adopt natural pest control techniques. Prevents harm to beneficial pollinators and their habitats.

10.2.4 Incorporate Agroforestry Practices:

Plant native trees pland shrubs along field boundaries or within the farm. Provides shelter, food, and nesting sites for pollinators. Examples: Flowering trees like neem, mango, or tamarind.

10.2.5 Use Cover Crops and Mulches:

Grow cover crops such as clover, buckwheat, or alfalfa that flower and attract pollinators. Enriches soil while offering pollinator forage.

10.2.6 Designate Pollinator Refuges:

Set aside undisturbed zones where natural vegetation can grow. Offers a safe haven for pollinators during farming activities.

10.2.7 Include Water Sources:

Provide shallow water sources with stones or floating plants for pollinators to drink safely. Ensures hydration for bees, butterflies, and other insects.

10.2.8 Promote Native Plant Species:

Use local flowering plants adapted to the region's climate and pollinators. Supports native pollinator species more effectively than exotic plants. Examples: Native wildflowers, herbs, or shrubs.

10.2.9 Practice Crop Intercropping:

Mix flowering plants with main crops to attract pollinators to the farm. Enhances pollination of the primary crop and provides a food source for pollinators. Examples: Intercrop marigolds, basil, or calendula with vegetables.

10.2.10 Avoid Monoculture:

Diversify cropping systems with multiple crop species. Attracts a wider variety of pollinators and reduces habitat disruption.

10.2.11 Promote Beneficial Insects:

Encourage natural predators like ladybugs and lacewings that cohabit with pollinators. Reduces competition and creates a balanced ecosystem.

10.2.12 Conduct Regular Monitoring:

Monitor pollinator activity and population diversity on the farm. Helps assess the effectiveness of conservation efforts and identify gaps.

10.2.13 Educate and Engage Farmers:

Train farmers on the importance of pollinators and their role in natural farming. Builds a community-driven approach to pollinator conservation.

10.2.14 Adopt Beekeeping:

Introduce managed honeybee colonies or foster native stingless bees (e.g., Trigona species). Farmers should avoid importing European honeybees (*Apis melliflora*) to establish new colonies in their natural farming agroecosystems. Instead, they should lure native honey bee species into their new bee boxes, from their environment, by using natural lures such as old beeswax or bee balm. In Japan and other Asian countries, oriental orchid (*Cymbidium floribundum*) plants are used to attract the Asian honey bee (*Apis cerana*), particularly *Apis cerana japonica* into new beeboxes. Boosts pollination efficiency while providing additional income through honey and wax production.

10.2.15 Time Framing Activities:

Schedule harvesting, and other disruptive activities during times when pollinators are less active (e.g., early morning or late evening). Reduces disturbance to pollinators.

10.3 Interspecific Interactions in Biotic Pollination of Crops:

Biotic pollination involves the transfer of pollen by living organisms such as insects, birds, bats, and other animals. Interspecific interactions in this context refer to the relationships between different species that influence pollination dynamics. These interactions can have significant implications for crop productivity and ecological balance.

Types of Interspecific Interactions in Biotic Pollination:

10.3.1 Mutualism:

Both the pollinator and the plant benefit. The pollinator gains nectar or pollen as a food source, while the plant achieves successful pollination. Example: Bees pollinating apple, almond, or mustard crops while collecting nectar.

10.3.2 Competition:

When multiple pollinator species or plants compete for the same resources, such as nectar, pollen, or pollinator services. Example: In a field with limited flowers, honeybees and native bees may compete for nectar.

10.3.3 Facilitation:

One species indirectly benefits another by enhancing its ability to attract pollinators. Example: A highly attractive flowering species in a mixed crop may increase pollinator visits to a less attractive neighboring crop.

10.3.4 Parasitism:

A pollinator or plant benefits at the expense of another species. Example: Nectar thieves (e.g., carpenter bees) that collect nectar without pollinating the flowers.

10.3.5 Predation:

Predators of pollinators can disrupt pollination. Example: Spiders or praying mantises preying on bees while they visit flowers.

CHAPTER 11

NATURAL CONTROL (ECOLOGICAL CONTROL) OF INSECT PESTS, DISEASES AND WEEDS

Conventional agriculture (monoculture or anti-biodiversity agriculture) uses many hateful terminologies for biodiversity such as weeds (unwanted plant biodiversity), pests (harmful insect biodiversity) and pathogens (causal organisms of plant diseases). In contrast, natural farming (polyculture or biodiversity based agriculture) considers these living organisms (biodiversity) as an integral part of the food web or food pyramid of the agroecosystem. For example, pests are the food of the predators and parasitoids, therefore the existence of pests are essential for survival of the predators and parasitoids in the agroecosystems. Farmers practicing natural farming, never get worried about pests, diseases or weeds.

11.1 Theories in Natural Control:

Pest-predator interactions are crucial for understanding ecological balance, pest control, and sustainable agriculture. Several theories and models have been developed to explain these interactions:

11.1.1 Lotka-Volterra Predator-Prey Model:

A mathematical model describing the dynamics of predator and prey populations in an ecosystem. Key Features: **Oscillations:** Predator and prey populations show cyclical changes over time. Predators increase when prey populations are high, leading to a subsequent decrease in prey, which in turn reduces predator numbers.

11.1.2 Functional Response Theory:

Explains how predator consumption rates change with prey density. Types: (1) **Type I**: Linear increase in predation until a maximum threshold (e.g., filter feeders). (2) **Type II**: Predation rate slows as prey density increases due to satiation (common in many predators). (3) **Type III**: Sigmoidal response where predators consume more prey at intermediate densities due to learning or prey switching.

11.1.3 Optimal Foraging Theory:

Predators maximize energy gain while minimizing effort. Key Assumptions: Predators prefer prey that offers the highest energy gain per unit effort. Search and handling times influence prey choice. Explains predator selectivity in pest management systems.

11.1.4 Enemy-Free Space Hypothesis:

Prey species exploit habitats or behaviors that reduce predation risk. Useful in designing cropping systems to minimize pest damage by providing refuges for natural predators.

11.1.5 Trophic Cascade Theory:

Predators indirectly regulate plant biomass by controlling herbivore populations. Example: Introducing a predator reduces pest populations, leading to less crop damage. Helps in understanding the cascading effects of predators in agroecosystems.

11.1.6 Ecological Stoichiometry:

Studies the balance of energy and nutrients between predators and prey. Nutritional content of pests can affect predator growth, reproduction, and efficiency. Guides the selection of predators in biocontrol programs.

11.1.7 Biocontrol Theory:

Focuses on the use of natural predators or parasitoids to control pest populations. Key Components: **Predator-prey ratios.** Timing and method of predator introduction. Widely used in Integrated Pest Management (IPM).

11.1.8 Interference Competition Theory:

Predicts how predators compete for limited prey. Helps explain predator efficiency in multi-predator systems. Relevant in agroecosystems with multiple natural enemies.

11.1.9 Evolutionary Arms Race:

Predators and prey continuously adapt to each other's strategies. Example: Pest species evolve resistance to predation, while predators develop improved hunting techniques. Explains the dynamics of pest resistance and predator adaptation.

11.1.10 Metapopulation Theory:

Examines predator and prey interactions in fragmented landscapes. Key Concept: Persistence depends on migration between patches. Relevant in designing landscape-level pest management strategies.

11.1.11 Island Biogeography Theory:

Predicts the diversity of species (including predators and pests) on islands based on size and isolation. Useful in understanding how habitat fragmentation affects pest and predator populations.

11.1.12 Density-Dependent Regulation:

Population growth rates of pests and predators are influenced by their densities. Key Concepts: **Positive Regulation:** High pest density attracts more predators. **Negative Regulation:** Overcrowding of predators reduces their efficiency. Helps optimize predator release strategies in pest control programs.

11.1.13 Natural Enemy Hypothesis:

States that ecosystems with higher biodiversity support more natural enemies of pests, reducing pest outbreaks. Promotes biodiversity conservation for pest management.

11.1.14 Niche Theory:

Explains how different species (pests and predators) coexist by utilizing distinct ecological niches. Key Ideas: Resource partitioning reduces competition. Predators with overlapping niches may compete for prey. Guides the selection of complementary predators for integrated pest management.

11.1.15 Janzen-Connell Hypothesis:

Suggests that pest pressure is higher near parent plants, limiting pest population growth and encouraging predator activity. Explains spatial distribution of pests and predators in agricultural fields.

11.1.16 Bottom-Up and Top-Down Control Theory:

Bottom-Up Control: Pest populations are regulated by resource availability (e.g., plant quality or quantity). **Top-Down Control:** Predators and parasitoids regulate pest populations. Balances resource management and predator introduction in crop systems.

11.1.17 Competition-Predation Trade-Off:

Explains how predators and pests coexist through trade-offs between competition for resources and predation risk. Helps design systems where pests face both competition and predation pressures.

11.1.18 Community Ecology Theory:

Studies interactions among multiple species, including predators, pests, and plants. Key Focus: Predator-prey dynamics within the broader food web. Helps manage complex agroecosystems with diverse species interactions.

11.1.19 Game Theory in Predator-Prey Interactions:

Analyzes strategic decision-making between predators and prey. Example: Prey may adopt hiding or mimicry strategies, while predators optimize hunting strategies. Predicts behavioral adaptations in pests and predators under varying conditions.

11.1.20 Life-Dinner Principle:

Predators and prey experience different evolutionary pressures because prey face higher stakes (life) than predators (a missed meal). Explains the often more rapid adaptation of pests compared to their predators.

11.1.21 Tragedy of the Commons:

In multi-predator systems, individual predators may overexploit prey, leading to population collapse. Guides sustainable predator-prey management to prevent pest overharvesting.

11.1.22 Herbivore Defense Theories:

Plant Defense Hypothesis: Plants evolve defenses (e.g., toxins or thorns) to deter pests, indirectly influencing predator-prey dynamics. **Enemy Release Hypothesis**: Pests become more problematic when introduced into areas without their natural predators. Guides pest management in native and invasive species scenarios.

11.1.23 Ecological Trap Theory:

Occurs when pests are attracted to suboptimal habitats due to altered environments, making them vulnerable to predators. Habitat management to create traps for pest populations.

11.1.24 Temporal Partitioning Theory:

Predators and prey adapt to different times of activity (e.g., diurnal vs. nocturnal) to reduce interaction intensity. Guides pest and predator activity monitoring in agriculture.

11.1.25 Resilience Theory:

Describes an ecosystem's ability to recover from disturbances affecting pest and predator populations. Helps in designing agricultural systems resilient to pest outbreaks.

11.1.26 Predator Swamping Theory:

Prey reduce predation risk by emerging in large numbers simultaneously, overwhelming predators. Explains pest population booms and predator response.

11.1.27 Host–Parasite Coevolution Theory:

Hosts (pests) and parasites evolve in response to each other's adaptations. Key Ideas: Hosts develop resistance mechanisms. Parasites evolve strategies to overcome host defenses. Explains the emergence of resistant pest populations and guides biocontrol efforts.

11.1.28 Virulence-Transmission Trade-Off Hypothesis:

Parasites must balance virulence (damage to the host) and transmission efficiency. Key Ideas: Highly virulent parasites may kill hosts before transmission. Less virulent parasites allow for longer host survival, aiding spread. Helps predict parasite effectiveness in pest population control.

11.1.29 Density-Dependent Parasitism:

The effectiveness of parasites increases with host density. Key Features: High pest densities attract more parasites, amplifying control. Low pest densities reduce parasite efficiency.

11.1.30 Red Queen Hypothesis:

Hosts and parasites must continuously adapt to survive in the evolutionary "arms race." Key Ideas: Parasites evolve to exploit pests. Pests develop resistance mechanisms, leading to continuous adaptation cycles. Explains pest resistance and parasite-host dynamics over time.

11.1.31 Host-Pathogen Dynamics Model:

Mathematical models (e.g., SIR model) describing the spread of parasites in pest populations. Key Parameters: **Susceptible** (S): Uninfected pests. **Infected** (I): Pests carrying parasites. **Recovered** (R): Pests that recover or die, reducing spread. Predicts the impact of parasites on pest populations.

11.1.32 Enemy Release Hypothesis:

Invasive pests often escape their natural parasites, leading to population explosions. Highlights the importance of introducing co-evolved parasites for invasive pest management.

11.1.33 Transmission Dynamics Theory:

Explores how parasites spread within pest populations. Key Factors: Contact rates between pests. Environmental conditions affecting parasite survival. Guides the deployment of parasitic agents under optimal conditions.

11.1.34 Niche Partitioning in Parasitism:

Different parasites exploit different niches within or among pest species. Reduces competition between parasites and enhances pest suppression. Useful for combining multiple parasitic agents.

11.1.35 Parasite-Mediated Competition:

Parasites reduce the competitive advantage of dominant pests, allowing subordinate species to thrive. Helps restore biodiversity in pest-dominated systems.

11.1.36 Pathogen-Host Specificity:

Parasites are often specific to certain pests, reducing non-target effects. High specificity makes parasites ideal for targeted biocontrol.

11.1.37 Allee Effect in Host-Parasite Systems:

Parasites may struggle to spread when pest densities are too low. Pest populations below a threshold are less vulnerable to parasitism. Explains the challenges of parasite establishment at low pest densities.

11.1.38 Immune Evasion Theory:

Parasites evolve mechanisms to evade pest immune responses. Parasitoid wasps injecting venom to suppress host immunity.

11.2 Methods of Natural Control Management in Natural Farming:

Natural control is an ecosystem service which is always present in the natural farming agroecosystem. Natural control is omnipresent and universal in the biosphere. Farmers can only manage the source of food and habitat of the organisms, to manage the population dynamics of insect pests, pathogens (causal organisms of diseases) and weeds in the natural farming agroecosystem. Here are the main methods used in natural control management:

11.2.1 Plant Breeding for Resistance:

Plant breeding for resistance, tolerance or immunity against insect pests, diseases, and environmental stresses is the most effective method for manipulating natural control mechanisms. This approach leverages genetic diversity and natural mechanisms to create sustainable and resilient agricultural systems. Key traits for natural control management: (1) **Pest and Disease Resistance**: (a) **Physical Barriers**: Thick leaf cuticles, hairy surfaces, or spines to deter herbivores. (b) **Chemical Defenses**: Production of secondary metabolites (e.g., alkaloids, tannins) that repel or kill pests. (c) **Hypersensitive Response**: Trigger localized cell death to stop pathogen spread. (2) **Competitive Growth**: Vigorous growth to outcompete weeds for light, nutrients, and water. (3) **Symbiotic Traits**: Ability to form beneficial associations with soil microbes (e.g., nitrogen-fixing bacteria, mycorrhizal fungi). (4) **Adaptive Root Systems**: Deep or fibrous roots for efficient nutrient and water uptake. (5) Stress Tolerance: Ability to withstand drought, salinity, or extreme temperatures.

11.2.2 No Till:

No-till farming is a conservation agriculture practice where soil is left undisturbed, and crops are planted directly into the residue of previous crops. This method promotes natural control of pests, diseases, and weeds by enhancing ecosystem balance and soil health. Here's how no-till farming helps: (A) **Natural Control of Pests**: (1) **Enhanced Biodiversity**: No-till systems create a stable habitat for beneficial predators, such as ground beetles, spiders, and parasitic wasps, which naturally control pest populations. Diverse soil microorganisms thrive, some of which are antagonistic to soil-borne pests and pathogens. (2) **Disruption of Pest Life Cycles**:

Leaving crop residues undisturbed disrupts pest habitats and prevents their breeding cycles. Rotation of crops in no-till systems reduces the buildup of host-specific pests. (3) **Balanced Ecosystem:** By maintaining a natural soil ecosystem, no-till farming reduces pest outbreaks often triggered by imbalances caused by tilling. (B) **Natural Control of Diseases:** (1) **Improved Soil Microbial Diversity:** No-till farming fosters beneficial soil microorganisms that suppress pathogenic fungi, bacteria, and nematodes. The competition between beneficial and harmful microbes helps maintain a healthy soil ecosystem. (2) **Reduced Soil Disturbance:** Tillage can bring dormant disease-causing pathogens to the surface. No-till farming minimizes this disturbance, reducing the risk of disease outbreaks. (3) **Crop Residue Management:** Crop residues left on the surface may initially harbor some pathogens but also support beneficial organisms that compete with or prey on these pathogens. (4) **Better Soil Structure:** Improved water infiltration and reduced compaction lower the likelihood of waterlogging, which can promote root diseases. (C) **Natural Control of Weeds:** (1) **Mulching Effect:** Crop residues act as a natural mulch, blocking sunlight and preventing weed seed germination. This reduces reliance on chemical herbicides. (2) **Reduced Seed Disturbance:** Tillage can bring buried weed seeds to the surface, where they germinate. No-till farming keeps weed seeds buried, reducing their chances of germination. (3) **Crop Competition:** No-till systems often involve cover cropping or intercropping, where living plants compete with weeds for resources like sunlight, water, and nutrients. (4) **Allelopathic Effects:** Certain cover crops (e.g., rye, clover) release chemicals into the soil that suppress weed germination and growth.

11.2.3 No Off-Farm Inputs:

Avoiding all off-farm inputs—such as chemical fertilizers, pesticides, herbicides, and synthetic amendments—can significantly contribute to the natural control of insect pests, diseases, and weeds by fostering a balanced and self-sustaining agricultural ecosystem. Here's how this approach supports natural control mechanisms: (A) **Natural Control of Insect Pests**: (1) **Promotion of Beneficial Predators**: Without chemical pesticides, natural predators (e.g., ladybugs, spiders, parasitic wasps) thrive and help control pest populations. Biodiversity in farming systems attracts pollinators and predator species that maintain pest balance. (2) **Enhanced Microbial Diversity**: Healthy soils, free of synthetic chemicals, support microbes that help suppress pest outbreaks and enhance plant immunity. (3) **Trap Cropping and Companion Planting**: Diversified planting systems (e.g., marigolds to repel nematodes) reduce pest populations without external inputs. (4) **Resilient Plants**: Plants grown in organic, nutrient-rich soil develop stronger natural defenses (e.g., thicker leaves, production of secondary metabolites) against pests. (B) **Natural Control of Diseases**: (1) **Improved Soil Health**: Avoiding synthetic inputs preserves beneficial soil microorganisms, which compete with pathogens and enhance soil immunity. Compost and organic matter addition encourage the growth of fungi and bacteria antagonistic to plant pathogens. (2) **Resilient Crops**: Crops grown in healthy soil are less prone to diseases due to stronger root systems and enhanced immunity. Disease outbreaks are minimized through crop rotation, diversity, and soil management practices. (3) **Reduced Chemical Stress**: Synthetic chemicals can weaken plants' natural defense mechanisms. Avoiding them allows plants to rely on and strengthen their inherent resistance. (C) **Natural Control of Weeds**: (1) **Mulching and Ground Cover**: Organic mulches suppress weed growth by blocking sunlight and creating a physical barrier. Living mulches (cover crops) compete with weeds for nutrients and space, reducing their establishment. (2) **Crop Competition**: Well-managed cropping systems with diverse plants suppress weeds by utilizing available resources efficiently. (3) **Natural Weed**

Suppression: Certain plants have allelopathic properties (e.g., rye, sunflower) that naturally suppress weed germination and growth. (4) **Soil Seed Bank Stability**: Tillage can disturb soil and bring buried weed seeds to the surface. No-till or minimal disturbance practices help keep weed seeds dormant.

11.2.4 No Irrigation:

Avoiding irrigation can help with the natural control of insect pests, diseases, and weeds by promoting a healthier, more balanced ecosystem. This approach relies on using water judiciously, often through rainfed agriculture, dryland farming, or techniques like mulching and moisture conservation to maintain soil and plant health. (A) **Natural Control of Insect Pests**: (1) **Reduced Favorable Conditions for Pests**: Over-irrigation creates moist environments that attract water-loving pests like aphids, thrips, and mites. Limiting irrigation reduces such conditions. Drier soils discourage the proliferation of soil-dwelling insect pests, such as root maggots and grubs. (2) **Improved Plant Resilience**: Avoiding excessive watering encourages plants to develop deeper, stronger roots, making them more resilient to pest attacks. Healthier plants grown under natural water regimes have better defenses against insect pests. (3) **Encouragement of Beneficial Insects**: A balanced soil moisture level supports beneficial insects, such as ground beetles and predatory mites, which help control pest populations. (B) **Natural Control of Diseases**: (1) **Prevention of Waterborne Diseases**: Excessive irrigation can lead to fungal and bacterial diseases like root rot, powdery mildew, and blight. Reducing irrigation minimizes waterlogging and prevents these diseases. (2) **Better Aeration and Soil Health**: Well-aerated soils, resulting from reduced irrigation, discourage anaerobic pathogens. Healthy soils support beneficial microbes that compete with and suppress disease-causing organisms. (3) **Less Disease Spread**: Irrigation methods like flooding or sprinklers can splash soil-borne pathogens onto plants. Avoiding irrigation reduces this risk. (C) **Natural Control of Weeds**: (1) **Weed Suppression through Water Limitation**: Many weeds thrive under conditions of excessive water. Limiting irrigation reduces their competitive advantage. Shallow-rooted weeds struggle to survive in dry soil conditions compared to deep-rooted crops. (2) **Encouraging Crop Competition**: By managing water carefully, crops can outcompete weeds, especially when combined with practices like mulching or cover cropping. (3) **Lower Weed Germination Rates**: Many weed seeds require specific moisture levels to germinate. Avoiding irrigation can disrupt their germination cycle.

11.2.5 Mulching:

Mulching is a sustainable agricultural practice where a layer of organic or inorganic material is applied to the soil surface. It plays a crucial role in the natural control of insect pests, diseases, and weeds by creating a protective and nurturing environment for plants. Here's how mulching contributes to pest, disease, and weed control: (A) **Natural Control of Insect Pests**: (1) **Disruption of Pest Life Cycles**: Mulch acts as a barrier, preventing soil-dwelling pests (e.g., cutworms, root maggots) from reaching the plant base. It disrupts the life cycles of pests by covering their breeding sites. (2) **Promotion of Beneficial Insects**: Organic mulch creates a habitat for predators like ground beetles and spiders, which feed on harmful pests. Earthworms and other decomposers thrive under mulch, indirectly enhancing plant health and resistance to pests. (3) **Moisture and Temperature Regulation**: By stabilizing soil temperature and moisture, mulch reduces plant stress, making plants less attractive to pests. (4) **Masking Crop Odors**: Certain pests locate plants by their odor. Mulch can mask these odors, reducing pest attraction. (B) **Natural Control of Diseases**: (1) **Reduction of Soil Splash**: Mulch prevents soil from splashing onto plants during irrigation or rain, which can spread soil-borne pathogens like fungal spores. (2) **Improved Soil Microbial Activity**: Organic mulch supports beneficial microbes and fungi that outcompete or inhibit disease-causing organisms. It promotes the growth of mycorrhizal fungi, which enhance plant resistance to root diseases. (3) **Decreased Humidity at the Plant Base**: By covering the soil, mulch reduces the humidity around plant stems, discouraging diseases like mildew and stem rot. (4) **Suppression of Pathogens**: Decomposing organic mulch releases substances like phenolics, which can suppress certain plant pathogens. (C) **Natural Control of Weeds**: (1) **Physical Barrier**: Mulch blocks sunlight from reaching weed seeds, preventing their germination and growth. This reduces the need for herbicides and manual weeding. (2] **Allelopathic Effects**: Certain mulches (e.g., straw, pine needles) release chemicals that inhibit weed seed germination and growth. (3) **Resource Competition**: Cover crops used as living mulch compete with weeds for water, light, and nutrients, suppressing their growth. (4) **Reduced Soil Disturbance**: Mulch eliminates the need for frequent soil tilling, which can bring buried weed seeds to the surface.

11.2.6 Cover Cropping:

Cover cropping is an effective natural strategy for controlling insect pests, diseases, and weeds in agricultural systems. Here's how it works: (1) **Control of Insect Pests**: (a) **Attraction of Beneficial Insects**: Cover crops like clover, buckwheat, and mustard attract predatory insects (e.g., ladybugs, parasitic wasps) and pollinators, which help control pest populations. (b) **Trap Crops**: Certain cover crops, such as mustard or radish, can act as trap crops by attracting pests away from the main crop. (c) **Habitat Provision**: Cover crops provide shelter for predators like spiders, ground beetles, and birds that feed on pests. (2) **Disease Suppression**: (a) **Biofumigation**: Cover crops from the Brassicaceae family (e.g., mustard, radish) release natural compounds (isothiocyanates) when decomposed, which suppress soil-borne pathogens and nematodes. (b) **Crop Rotation Benefits**: Using cover crops disrupts the life cycle of disease-causing organisms by creating a break in host availability. (c) **Improved Soil Health**: Cover crops increase organic matter and microbial diversity, which can outcompete pathogens and enhance plant immunity. (3) **Weed Suppression**: (a) **Canopy Cover**: Dense-growing cover crops like rye, oats, or hairy vetch shade the soil, preventing sunlight from reaching weed seeds and reducing germination. (b) **Allelopathy**: Some cover crops (e.g., rye, sorghum) release chemicals that inhibit weed seed germination and growth. (c) **Physical Barrier**: A thick layer of cover crop residue (green mulch) suppresses weeds by creating a physical barrier on the soil surface.

11.2.7 Multilayer Polycropping:

Multilayer polycropping systems are highly effective in naturally controlling insect pests, diseases, and weeds due to their diverse and synergistic interactions among plants. Here's how these systems work: (1) **Control of Insect Pests**: (a) **Disruption of Pest Habitats**: Diverse plant layers (trees, shrubs, herbs, and ground covers) create complex habitats that make it harder for pests to locate their preferred host plants. (b) **Natural Predators and Parasitoids**: Multilayer systems attract and sustain beneficial insects like ladybugs, spiders, and parasitic wasps by providing food and shelter. (c) **Trap Cropping**: Certain layers can act as trap crops, attracting pests away from primary crops. For example, marigolds in the lower layer can repel nematodes and attract aphids, diverting them from vegetables. (d) **Pest Deterrence**: Companion plants with pest-repelling properties, like garlic, basil, or mint, can be included in different layers to deter specific pests. (2) **Disease Suppression**: (a) **Microclimate Regulation**: Multilayer systems reduce humidity and temperature fluctuations by providing shade and airflow, minimizing the conditions that favor disease proliferation. (b)

Barrier Effect: Upper layers, such as tall trees or shrubs, act as physical barriers, preventing the spread of airborne pathogens to lower layers. (c) **Disease Breakthrough Prevention:** Crop diversity reduces the risk of disease spreading, as pathogens often target specific plant species. For instance, intercropping legumes with cereals can disrupt fungal diseases. (3) **Weed Suppression:** (a) **Shading and Smothering:** The dense canopy of multilayer crops shades the soil, preventing sunlight from reaching weed seeds and suppressing their growth. Ground cover crops like sweet potato or pumpkin smother weeds. (b) **Root Competition:** Diverse root systems compete with weeds for nutrients, water, and space, leaving little opportunity for weeds to thrive. (c) **Mulch from Organic Matter:** Fallen leaves and crop residues from upper layers create a natural mulch that suppresses weed germination and growth.

11.2.8 Carnivorous Plants:

Carnivorous plants can aid in the natural control of insect pests by attracting, capturing, and digesting them, thereby reducing their populations in specific environments (Axel Mithofer 2022). Here are some examples: (1) **Venus Flytrap** (*Dionaea muscipula*): Possesses modified leaves that snap shut when trigger hairs are stimulated by unsuspecting insects, effectively trapping and digesting them. While primarily grown for ornamental purposes, Venus flytraps can help manage indoor insect pests like flies and gnats. (2) **Pitcher Plants** (*Sarracenia* spp. and *Nepenthes* spp.): Feature deep cavity traps filled with digestive fluids. Insects are lured by nectar and coloration, fall into the cavity, and are unable to escape. Pitcher plants can capture a variety of insects, including flies, ants, and beetles. Studies have explored their potential in controlling specific pests; for instance, research on Sarracenia species indicated that while they do capture insects, they may not significantly reduce pest populations in agricultural settings. (Mariyappillai A et al. 2019). (3) **Sundews** (*Drosera* spp.): Covered with glandular hairs that secrete sticky mucilage, sundews trap insects on their leaf surfaces and slowly digest them. Sundews can manage small insect pests like gnats and aphids in controlled environments. (4) **Bladderworts** (*Utricularia* spp.): Aquatic or semi-aquatic plants with specialized bladder-like traps that create a vacuum to suck in small prey, including insect larvae. Research has shown that certain bladderwort species can reduce mosquito larvae populations in water bodies, indicating potential for natural mosquito control (Mohanty AK et al. 2024)..

11.3 Interspecific Interactions in Natural Control of Pests, Pathogens and Weeds:

Plant Defense against Insect Pests, Diseases and Weeds:

Plants have evolved a variety of defense mechanisms to protect themselves from insect pests and diseases. These defenses can be physical, chemical, or biological in nature. Here are some examples: (1) **Physical Defenses:** (a) **Thorns and Spines:** Plants like roses and cacti use thorns and spines to deter herbivores and insects. (b) **Trichomes:** Hair-like structures on leaves and stems (e.g., in tomatoes and cotton) can trap or repel insects. (c) **Thick Bark and Waxy Cuticles:** Trees like oak have thick bark to resist pests, while plants like cabbage have waxy layers to prevent fungal infections. (d) **Leaf Toughness:** Plants like holly have tough, spiny leaves that are difficult for insects to chew. (2) **Chemical Defenses:** (a) **Toxic Compounds: Alkaloids:** Nicotine in tobacco plants is toxic to insects. **Tannins:** Oaks produce tannins that deter herbivores by making the plant unpalatable. **Cyanogenic Glycosides:** Cassava releases hydrogen cyanide when attacked. (b) **Repellents:** Essential oils in plants like lavender and citronella repel insects. (b) **Phytoalexins:** Grapevines produce these antimicrobial compounds to combat fungal infections. (c) **Protease Inhibitors:** Found in legumes like soybeans, these disrupt insect digestion. (3)

Biological Defenses: (a) **Mutualistic Relationships: Ant–Plant Symbiosis:** Acacia trees provide food and shelter for ants, which in turn protect the tree from herbivores. **Predator Attraction:** Some plants, like maize, release volatile organic compounds (VOCs) to attract predatory insects like parasitic wasps that attack herbivores. (4) **Induced Defenses:** (a) **Systemic Acquired Resistance (SAR):** When attacked, plants like tobacco produce salicylic acid, which triggers defense responses throughout the plant. (b) **Jasmonic Acid Pathway:** In response to chewing insects, plants like Arabidopsis increase production of jasmonic acid to fortify defenses. (c) **Hypersensitive Response (HR):** Plants like tobacco and potato trigger localized cell death around infection sites to stop pathogens from spreading. (5) **Mimicry and Camouflage:** (a) **Leaf Mimicry:** Passionflower vines mimic the appearance of butterfly eggs on their leaves to deter egg-laying. (b) **Crypsis:** Certain plants, like Lithops (stone plants), mimic rocks to avoid detection. (6) **Allelopathy:** Plants like marigold release chemicals into the soil that suppress nematodes and soil-borne pathogens.

Insect Predators Control Insect Pests:

Insect predators play a vital role in controlling pest populations naturally by feeding on them. Here are examples of key insect predators and the pests they target: (1) **Ladybugs** (Ladybird Beetles): *Coccinellidae* spp. Target Pests: Aphids, Whiteflies, Mealybugs, Spider mites. Ladybugs can significantly reduce aphid populations on crops like roses, tomatoes, and peppers. (2) **Lacewings:** *Chrysopidae* spp. Target Pests: Aphids, Thrips, Whiteflies, Caterpillars. Green lacewing larvae, known as "aphid lions," devour aphids on lettuce and other leafy greens. (3) **Praying Mantises:** *Mantodea* spp. Target Pests: Grasshoppers, Caterpillars, Moths, Beetles. Praying mantises control caterpillar populations in vegetable gardens like cabbage and broccoli. (4) **Ground Beetles:** *Carabidae* spp. Target

Pests: Slugs, Caterpillars, Root maggots, Cutworms. Ground beetles protect crops like corn and potatoes by preying on root-feeding pests. (5) **Hoverflies** (Syrphid Flies): *Syrphidae* spp. Target Pests: Aphids, Scales, Mealybugs. Hoverfly larvae feed on aphids in flower gardens and vegetable fields. (6) **Assassin Bugs**: *Reduviidae* spp. Target Pests: Leafhoppers, Caterpillars, Aphids. Assassin bugs help manage pest populations in orchards and vineyards. (7) **Spiders**: Various species (e.g., **Araneae** spp.). Target Pests: Mosquitoes, Flies, Moths, Beetles. Orb-weaver spiders trap flying pests like moths and flies in their webs in agricultural fields. (8) **Parasitic Wasps**: Various species (e.g., *Trichogramma* spp., *Aphidius* spp.). Target Pests: Caterpillars (e.g., corn earworm), Whiteflies, Aphids, Leaf miner. Trichogramma wasps parasitize the eggs of moths and butterflies, preventing infestations in cornfields and fruit orchards. (9) **Dragonflies and Damselflies**: *Odonata* spp. Target Pests: Mosquitoes, Gnats, Flies. Dragonflies control mosquito populations in wetlands and rice paddies. (10) **Ants**: Various species (e.g., *Formicidae* spp.). Target Pests: Termites, Caterpillars, Eggs of various pests. Weaver ants protect citrus crops by preying on caterpillars and beetles. (11) **Predatory Stink Bugs**: *Podisus* spp. Target Pests: Caterpillars, Beetle larvae, Leafhoppers. Predatory stink bugs are used in cotton fields to control bollworm larvae. (12) **Predatory Mites**: *Phytoseiidae* spp. Target Pests: Spider mites, Thrips, Whiteflies. Predatory mites manage spider mite infestations in greenhouses and fruit crops.

Insect Parasitoids Control Insect Pests:

Insect parasitoids are natural biological control agents that help manage pest populations. They lay their eggs in or on a host insect, and the developing larvae eventually kill the host. Here are key examples of insect parasitoids and the pests they control: (1) **Parasitic Wasps**: (a) *Trichogramma* spp. Target Pests: Moth and butterfly eggs (e.g., corn earworm, cotton bollworm). The wasp lays its eggs inside the pest eggs, preventing them from hatching. (b) *Aphidius* spp. Target Pests: Aphids (e.g., green peach aphid, cotton aphid). The wasp lays eggs inside the aphid, leading to the aphid's death as the larvae develop. (c) *Encarsia formosa*: Target Pests: Whiteflies (e.g., greenhouse whitefly). The wasp parasitizes whitefly pupae, reducing pest numbers. (d) *Cotesia* spp.: Target Pests: Caterpillars (e.g., armyworms, diamondback moth). The wasp injects eggs into caterpillars, and larvae feed internally, eventually killing the host. (2) **Tachinid Flies** (*Tachinidae* spp.): Target Pests: Caterpillars, beetles, and grasshoppers. Tachinid flies are effective against pests like gypsy moth larvae and squash bugs. Females lay eggs on or near the host, and the larvae burrow into the pest to feed internally. (3) **Braconid Wasps** (*Braconidae* spp.): Target Pests: Aphids, caterpillars (e.g., tomato hornworm), and beetle larvae. Braconid wasps parasitize tomato hornworms, leaving behind parasitized worms with visible white cocoons. Injects eggs into the host; larvae consume internal tissues, killing the pest. (4) **Ichneumonid Wasps** (Ichneumonidae spp.): Target Pests: Wood borers, caterpillars, and sawfly larvae. Lays eggs inside the pest's body; larvae develop and kill the host. (5) **Predatory Parasitic Flies** (Conopidae spp.): Target Pests: Grasshoppers, beetles, and caterpillars. (6) **Strepsiptera** (Twisted-Wing Parasites): Target Pests: Leafhoppers, planthoppers, and wasps. The parasite lives inside the host, affecting its reproduction and eventually killing it. (7) **Nematodes** as Parasitoids (*Steinernema* spp., *Heterorhabditis* spp.): Target Pests: Soil-dwelling pests like root maggots, grubs, and weevils. Parasitic nematodes infect the pest and release symbiotic bacteria that kill the host.

Entomopathogenic Fungi Control Insect Pests:

Entomopathogenic fungi are natural pathogens of insects that infect and kill pests, playing a significant role in biological pest control. They invade insects through their cuticle, grow internally, and ultimately kill the host. (1) **Beauveria bassiana**: Target Pests: Aphids, Whiteflies, Thrips, Grasshoppers, Termites, Colorado potato beetles. Spores attach to the insect's cuticle, penetrate the body, and grow internally, killing the pest within days. (2) **Metarhizium anisopliae**: Target Pests: Termites, Root weevils, Locusts, Mosquito larvae, Beetles. Spores germinate on the pest's body, invade through the cuticle, and release toxins that kill the insect. (3) **Lecanicillium lecanii**: Target Pests: Aphids, Whiteflies, Thrips, Scale insects. Spores infect the insect's body, disrupt its feeding, and lead to death. (4] **Hirsutella thompsonii**: Target Pests: Mites (e.g., citrus rust mite, spider mites). Fungus parasitizes the mites, reducing their population significantly. (5) **Isaria fumosorosea** (*Paecilomyces fumosoroseus*): Target Pests: Whiteflies, Thrips, Aphids, Mealybugs. Fungal spores adhere to the insect's body, penetrate, and kill it within a few days. (6) **Cordyceps** spp. (e.g., *Cordyceps militaris*): Target Pests: Caterpillars, Grasshoppers. Some species naturally parasitize forest pests like gypsy moth larvae. Spores infect the host, take over its body, and emerge to release new spores. (7) **Nomuraea rileyi**: Target Pests: Caterpillars (e.g., armyworms, diamondback moth). Effective against fall armyworm in maize crops. The fungus invades the caterpillar's body, leading to death within a week. (8) **Pandora neoaphidis**: Target Pests: Aphids. Found naturally infecting aphids in wheat and vegetable fields. Spores spread through the air, infect aphids, and kill them rapidly.

Bacterial Pathogens Control Insect Pests:

Bacteria can act as effective biological control agents against insect pests by producing toxins, causing diseases, or interfering with the pests' life cycle. (1) **Bacillus thuringiensis** (Bt): Target Pests: Caterpillars (e.g., corn earworm, cabbage looper, diamondback moth), Mosquito larvae, Beetles (e.g., Colorado potato beetle). Produces crystal (Cry) proteins toxic to specific insect larvae. When ingested, the toxins disrupt the insect's gut, causing death. (2) **Bacillus sphaericus**: Target Pests: Mosquito larvae. Produces toxins that specifically target mosquito larvae in water bodies. (3) **Pseudomonas fluorescens**: Target Pests: Root-knot nematodes, Soil-dwelling insect pests. Produces antimicrobial compounds and suppresses the growth of pests by colonizing plant roots. (4) **Serratia entomophila**: Target Pests: Grass grubs (e.g., *Costelytra zealandica*). Causes amber disease, which leads to starvation and death in grass grub larvae. (5) **Lysinibacillus sphaericus**: Target Pests: Mosquito larvae. The bacterium's spores and toxins are ingested by mosquito larvae, killing them through gut disruption. Effective in controlling Anopheles and Culex mosquito species in stagnant water. (6) **Paenibacillus popilliae**: Target Pests: Japanese beetle larvae (Popillia japonica). Causes milky spore disease, killing the larvae by infecting their digestive system. (7) **Wolbachia** spp.: Target Pests: Mosquitoes (*Aedes aegypti*), fruit flies, and other arthropods. Alters reproductive capabilities by inducing cytoplasmic incompatibility or reducing the pest's ability to transmit diseases (e.g., dengue, Zika). (8) **Burkholderia** spp.: Target Pests: Root-damaging pests. Produces insecticidal compounds that kill soil-dwelling insect pests. (9) **Chromobacterium subtsugae**: Target Pests: Aphids, Whiteflies, Thrips, Beetle larvae. Produces metabolites toxic to pests while being safe for non-target organisms. (10) **Saccharopolyspora spinosa**: Target Pests: Caterpillars, Thrips, Leafminers. Produces spinosyns, which are neurotoxins that paralyze and kill pests.

Insect Pathogenic Viruses Control Insect Pests:

Certain viruses are natural biological control agents of insect pests of crops. Most belong to the family Baculoviridae, which specifically target insects without harming plants, humans, or other animals.

(1) **Nucleopolyhedrovirus (NPV)**: Target Pests: Caterpillars (e.g., armyworms, cotton bollworms, diamondback moths, fall armyworms). NPV infects and multiplies inside the insect's body, disrupting cellular functions and leading to death. Infected insects appear liquefied or discolored. (2) **Granulovirus (GV)**: Target Pests: Beetles (e.g., codling moth (Cydia pomonella), potato tuber moth, gypsy moth larvae). The virus infects the midgut of larvae after ingestion, leading to starvation and death. (3) **Autographa californica NPV (AcMNPV)**: Target Pests: Lepidopteran pests (e.g., loopers, cabbage worms). Infects caterpillars, leading to systemic infection and death within days. (4) **Spodoptera frugiperda NPV (SfNPV)**: Target Pests: Fall armyworm (Spodoptera frugiperda). Causes viral disease in larvae, reducing pest populations. (5) **Oryctes rhinoceros Virus (OrV)**: Target Pests: Coconut rhinoceros beetle (*Oryctes rhinoceros*). Infects adult beetles, reducing their reproduction and causing death. (6) **Densoviruses (DNV)**: Target Pests: Mosquito larvae, Lepidopteran larvae. Infects insect gut cells, causing paralysis and death. (7) **Hytrosaviruses**: Target Pests: Tsetse flies (Glossina spp.). Causes reproductive disorders in tsetse flies, reducing population growth. (8) **Entomopoxvirus**: Target Pests: Grasshoppers, Locusts, Lepidopteran pests. Infects and disrupts the digestive and reproductive systems of the pest. (9) **Reoviruses (Cypoviruses)**: Target Pests: Lepidopteran pests (e.g., silkworms, gypsy moths). Targets the midgut of larvae, causing stunted growth and death. (10) **Iridoviruses**: Target Pests: Beetles, Caterpillars, Grasshoppers. Causes iridescent coloring in infected larvae due to viral replication and eventual death.

Fungal Antagonists Control Fungal Pathogens:

Fungi that control fungal pathogens in crops are an integral part of biological control strategies. These fungi work through mechanisms like mycoparasitism, competition, or the production of antifungal compounds. (1) **Trichoderma** spp.: Target Pathogens: *Rhizoctonia solani* (root rot), *Sclerotinia sclerotiorum* (stem rot), *Fusarium oxysporum* (wilt), *Pythium* spp. (damping-off), *Botrytis cinerea* (gray mold). Mycoparasitism: Trichoderma directly attacks and degrades the cell walls of pathogenic fungi. Produces antifungal metabolites and competes for nutrients and space. (2) **Gliocladium** spp.: Target Pathogens: *Fusarium* spp. (wilt diseases), *Pythium* spp. (damping-off), *Phytophthora* spp. (blight). Produces enzymes that degrade pathogenic fungal cell walls. Acts as a competitive antagonist by occupying space and utilizing resources. (3) **Coniothyrium minitans:** Target Pathogens: *Sclerotinia sclerotiorum* (stem rot in beans, sunflowers, and canola), *Botrytis cinerea* (gray mold in fruits and vegetables). Mycoparasitism: Attacks and kills sclerotia (resting structures) of pathogenic fungi. Reduces disease recurrence by depleting pathogen reservoirs. (4) **Ampelomyces quisqualis:** Target Pathogens: Powdery mildew fungi (*Erysiphe* spp., *Oidium* spp.). Mycoparasitism: Invades the hyphae and fruiting bodies of powdery mildew fungi, reducing their growth and sporulation. (5) **Chaetomium globosum:** Target Pathogens: *Fusarium* spp. (wilt diseases), *Pythium* spp. (damping-off), *Rhizoctonia solani* (root rot). Produces antifungal compounds like chaetoglobosin. Competes with pathogens for nutrients and space. (6) **Clonostachys rosea** (formerly *Gliocladium roseum*): Target Pathogens: *Botrytis cinerea* (gray mold), *Sclerotinia* spp. (stem rot). Mycoparasitism: Invades and destroys fungal pathogens' structures. Produces antifungal secondary metabolites. (7) **Pythium oligandrum:** Target Pathogens: *Pythium* spp. (damping-off), *Phytophthora* spp. (late blight). Mycoparasitism: Parasitizes other Pythium and Phytophthora species. Induces systemic resistance in plants. (8) **Beauveria bassiana:** Target Pathogens: Fungal pathogens causing wilt and root rot diseases. Produces enzymes and antifungal compounds that degrade fungal pathogens. (9) **Verticillium lecanii:** Target Pathogens: *Botrytis cinerea* (gray mold), *Sclerotinia* spp. (stem rot). Invades and suppresses fungal growth through parasitism and toxin production. (10) **Aspergillus niger**: Target Pathogens: Post-harvest fungal pathogens (*Penicillium* spp., *Fusarium* spp.). Produces antifungal metabolites that inhibit the growth of spoilage fungi.

Bacterial Antagonists Control Bacterial Pathogens:

Certain bacteria are effective biological control agents against bacterial plant pathogens. These beneficial bacteria work through competition, production of antimicrobial compounds, induction of systemic resistance in plants, or direct inhibition of pathogens. (1) **Pseudomonas fluorescens**: Target Pathogens: *Xanthomonas campestris* (black rot in cabbage), *Ralstonia solanacearum* (bacterial wilt in tomatoes and potatoes), *Erwinia amylovora* (fire blight in apple and pear). Produces antibiotics like phenazines, pyoluteorin, and 2,4-diacetylphloroglucinol (DAPG) that inhibit pathogens. Competes with pathogens for nutrients and colonization sites. Induces systemic resistance in plants. (2) **Bacillus subtilis**: Target Pathogens: *Xanthomonas* spp. (leaf spots and blights), *Pseudomonas syringae* (bacterial speck), *Erwinia carotovora* (soft rot). Produces antimicrobial compounds like iturins, fengycins, and surfactins. Forms biofilms that exclude pathogens. Induces plant defense mechanisms. (3) **Bacillus amyloliquefaciens**: Target Pathogens: *Xanthomonas oryzae* (bacterial blight in rice), *Pseudomonas syringae* (bacterial speck in tomatoes), *Ralstonia solanacearum* (bacterial wilt). Produces antimicrobial lipopeptides like surfactin and fengycin. Competes for nutrients and iron (via siderophores). (4) **Agrobacterium radiobacter**: Target Pathogens: *Agrobacterium tumefaciens* (crown gall disease). Produces bacteriocin (Agrocin 84) that specifically inhibits A. tumefaciens. (5) **Streptomyces** spp.: Target Pathogens: *Erwinia* spp. (soft rot), *Pseudomonas syringae* (bacterial speck), *Ralstonia* spp. (wilt diseases). Produces antibiotics like streptomycin, actinomycin, and tetracycline. Competes for resources and induces plant systemic resistance. (6) **Lysinibacillus sphaericus:** Target Pathogens: *Xanthomonas* spp. (leaf blight in cereals), *Pectobacterium carotovorum* (soft rot). Produces extracellular enzymes and antimicrobial peptides that disrupt bacterial pathogens. (7) **Burkholderia cepacia:** Target Pathogens: *Ralstonia solanacearum* (bacterial wilt), *Xanthomonas* spp. (leaf blight). Produces antifungal and antibacterial metabolites. Induces systemic resistance in plants. (8) **Paenibacillus polymyxa:** Target Pathogens: *Ralstonia solanacearum* (bacterial wilt), *Xanthomonas campestris* (black rot). Produces antibiotics like fusaricidins and polymyxins. Promotes plant growth and enhances disease resistance. (9) **Pseudomonas putida:** Target Pathogens: *Pseudomonas syringae* (bacterial speck), *Xanthomonas* spp. (leaf spot and blight diseases). Produces siderophores, lytic enzymes, and antimicrobial compounds. Competitively colonizes roots and leaves. (10) **Ralstonia eutropha:** Target Pathogens: Soil-borne bacterial pathogens like *Ralstonia solanacearum*. Produces bacteriocins that inhibit pathogenic bacterial strains.

Fungal Pathogens Control Weeds:

Certain fungi, known as **mycoherbicides**, are used as biological control agents to manage weeds in agriculture. These fungi specifically target and suppress weed growth through infection, competition, or production of toxins, offering an environmentally friendly alternative to chemical herbicides. (1) **Colletotrichum gloeosporioides:** Target Weeds: Tropical soda apple (*Solanum viarum*), Northern jointvetch (*Aeschynomene virginica*). Causes anthracnose disease in weeds, leading to leaf lesions, defoliation, and plant death. (2) **Alternaria alternata:** Target Weeds: Water hyacinth (*Eichhornia crassipes*), Velvetleaf (*Abutilon theophrasti*). Produces toxins that inhibit weed growth and cause leaf blight. (3) **Puccinia chondrillina:** Target Weeds: Skeleton weed (*Chondrilla juncea*). A rust fungus that infects leaves and stems, reducing photosynthesis and seed production. (4) **Fusarium oxysporum** (Pathotype-Specific Strains): Target Weeds: Striga (witchweed) species (*Striga hermonthica* and *Striga asiatica*). Infects the roots of parasitic weeds, disrupting their attachment to host plants. (5) **Phytophthora palmivora**: Target Weeds: Milkweed vine (*Morrenia odorata*). Causes root and stem rot in weeds, leading to plant death. (6) **Sclerotinia minor**: Target Weeds: Dandelion (*Taraxacum officinale*), Canada thistle (*Cirsium arvense*). Infects and degrades the roots and basal leaves of weeds, causing wilting and death. (7) **Cercospora rodmanii**: Target Weeds: Water hyacinth (*Eichhornia crassipes*). Causes leaf spot disease, reducing weed vigor and reproductive capacity. (8) **Uromyces gallii–vernicis:** Target Weeds: Gorse (*Ulex europaeus*). A rust fungus that infects and weakens gorse plants by damaging their leaves and stems. (9) **Epicoccum purpurascens:** Target Weeds: Dodder (*Cuscuta* spp.). Infects dodder seeds and seedlings, reducing their germination and growth. (10) **Cochliobolus lunatus:** Target Weeds: Barnyard grass (*Echinochloa crus-galli*). Produces toxins that inhibit seedling growth and cause plant death.

Bacterial Pathogens Control Weeds:

Certain bacteria act as biological control agents for weeds by producing toxins, suppressing weed seed germination, or inhibiting growth. These bacteria offer an environmentally friendly alternative to chemical herbicides. (1) **Pseudomonas fluorescens:** Target Weeds: Downy brome (*Bromus tectorum*), Jointed goatgrass (*Aegilops cylindrica*). Produces allelopathic compounds that inhibit seed germination and root elongation. Competes with weed-associated microorganisms, suppressing weed growth. (2) **Xanthomonas campestris** pv. **poae:** Target Weeds: Annual bluegrass (*Poa annua*). Infects and causes blight in bluegrass, leading to the death of the weed. (3) **Burkholderia cepacia:** Target Weeds: Canada thistle

(*Cirsium arvense*). Produces phytotoxins that damage the roots and shoots of the weed. Colonizes weed rhizospheres, outcompeting other soil microbes beneficial to the weed. (4) **Pseudomonas syringae** pv. **tagetis:** Target Weeds: Canada thistle (*Cirsium arvense*), Common ragweed (*Ambrosia artemisiifolia*). Produces a toxin called tagetitoxin that inhibits chloroplast development, causing chlorosis and stunting weed growth. (5) **Ralstonia solanacearum** (Weed-Specific Strains): Target Weeds: Water hyacinth (*Eichhornia crassipes*). Infects and blocks vascular tissues, causing wilting and death. (6) **Streptomyces** spp.: Target Weeds: Parasitic weeds like Striga (witchweed). Produces secondary metabolites that inhibit the germination of parasitic weed seeds. (7) **Bacillus subtilis:** Target Weeds: Invasive grasses and broadleaf weeds. Produces antifungal and antibacterial compounds that suppress weed seed germination and growth. (8) **Clavibacter toxicus:** Target Weeds: Annual ryegrass (*Lolium rigidum*). Produces toxic metabolites that inhibit weed growth and development. (9) **Azospirillum brasilense:** Target Weeds: Grassy weeds in cereal crops. Competes with weed-associated microorganisms and alters soil nutrient dynamics, suppressing weed growth. (10) **Mycobacterium phlei:** Target Weeds: Creeping thistle (*Cirsium arvense*). Produces phytotoxic compounds that inhibit weed shoot and root growth.

11.4 Biomonitoring of Natural Control:

Biomonitoring of natural pest control involves tracking and assessing the presence and activity of beneficial organisms that regulate pest populations in agricultural ecosystems. Here are some methods used in biomonitoring natural control of insect pests:

Field Observation and Sampling: Visual Monitoring: Regular field visits to observe beneficial organisms (e.g., predators, parasitoids, and pollinators) and pest interactions. Look for evidence of predation (e.g., damaged pest bodies) or parasitism (e.g., parasitized pest larvae or pupae). **Sweep Net Sampling**: Use a sweep net to capture insects from crops and surrounding vegetation. Identify and count natural enemies like ladybirds, lacewings, and predatory beetles. **Pitfall Traps**: Place pitfall traps to monitor ground-dwelling predators such as ants, spiders, and carabid beetles. **Sticky Traps**: Deploy sticky traps to capture flying insects, including parasitoids and predatory wasps.

Assessing Parasitism and Predation: Parasitism Rate: Collect pest samples (e.g., caterpillars or aphids) and rear them in controlled conditions to identify emerging parasitoids. Calculate parasitism rates to measure the effectiveness of parasitoids like Trichogramma wasps or tachinid flies. **Predation Rate**: Use sentinel prey (e.g., pest eggs or larvae) placed in the field to assess predation levels. Record the proportion of prey removed or consumed over a set time.

Population Dynamics Modeling: Pest-to-Predator Ratios: Calculate the ratios of pest populations to natural enemy populations. A lower ratio indicates effective natural pest control. **Population Tracking**: Use time-series data on pest and natural enemy populations to model interactions and predict control effectiveness.

Biodiversity Assessment: Species Richness and Abundance: Assess the diversity and abundance of natural enemies in the crop ecosystem. High biodiversity often correlates with better pest control. **Indicator Species**: Identify key indicator species that reflect the health of the pest control ecosystem, such as specific predatory beetles or parasitoid wasps.

Crop Damage Assessment: Damage Surveys: Examine crop plants for pest damage and correlate with the presence of natural enemies. Reduced damage in areas with high natural enemy activity indicates effective control. **Yield Analysis**: Compare yields in areas with active biomonitoring of natural enemies to those with chemical pest control.

Artificial Refugia and Habitat Manipulation: Monitoring in Refugia: Install artificial habitats (e.g., insect hotels or hedgerows) to attract and monitor natural enemies. Track the colonization and pest control activity in these habitats. **Floral Resources:** Plant flowering strips to attract pollinators and parasitoids. Monitor their impact on pest populations in adjacent crops.

Farmer Participation: Citizen Science: Engage farmers in recording sightings of natural enemies and pests. Use mobile apps or logbooks to gather data from multiple fields. **Pheromone Traps:** Set up pheromone traps for pests and monitor the impact of natural enemies on trap captures.

Long-term Monitoring Programs: Establish permanent monitoring plots to track natural pest control trends over multiple seasons. Collect data on climatic conditions, crop rotation, and pesticide use to correlate with natural pest control effectiveness.

CHAPTER 12

INTEGRATED ANIMAL HUSBANDRY

12.1 Objectives of Integrating Animal Husbandry Natural Farming:

Integrating animal husbandry with natural farming agribusiness aligns with the principles of sustainability, biodiversity, and holistic farm management. Below are the primary objectives of such integration:

12.1.1 Enhanced Soil Fertility and Health:

Use animal manure and by-products to enhance soil fertility naturally. Improves soil structure, microbial activity, and nutrient availability for crops.

12.1.2 Waste Recycling and Zero Waste:

Utilize agricultural residues as animal feed and recycle animal waste as organic compost or biogas slurry. Creates a closed-loop system with no waste, minimizing environmental pollution.

12.1.3 Diversified Income Streams:

Generate multiple income sources from crop production, dairy, poultry, fish farming, or other animal products. Reduces financial risks and increases overall farm profitability.

12.1.4 Sustainable Resource Utilization:

Optimize the use of available land, water, and biological resources. Improves resource efficiency and ensures long-term farm viability.

12.1.5 *Improved Farm Productivity:*

Increase overall farm output by utilizing synergies between crops and livestock. Higher yield per unit of land and enhanced resource productivity.

12.1.6 *Reduction in Input Costs:*

Lower dependence on expensive synthetic inputs by using organic fertilizers, natural pest control, and locally available resources. Makes farming more economical and sustainable.

12.1.7 *Climate Resilience:*

Build a resilient farming system capable of withstanding climate variability through diverse practices. Reduces risks from crop failures or livestock losses during extreme weather events.

12.1.8 *Biodiversity Conservation:*

Foster biodiversity by integrating livestock, crops, trees, and natural habitats. Promotes ecological balance and protects native species.

12.1.9 *Natural Pest and Weed Control:*

Use grazing animals or poultry to control weeds and pests, reducing the need for chemicals. Maintains a healthy ecosystem and minimizes environmental damage.

12.1.10 *Employment Generation:*

Create opportunities for skilled and unskilled labor in activities like milking, composting, feeding, and processing. Improves rural livelihoods and reduces migration to urban areas.

12.1.11 *Value Addition:*

Add value to farm products like milk, eggs, and manure through processing and branding. Increases profit margins and market competitiveness.

12.1.12 Self-Sufficiency and Food Security:

Ensure farm families have access to diverse and nutritious food from crops and livestock. Promotes food security at the household and community levels.

12.1.13 Promotion of Ethical and Eco-Friendly Practices:

Encourage humane treatment of animals and environmentally sound practices. Builds consumer trust and attracts markets for ethical, organic produce.

12.2 Strategies for Integrated Animal Husbandry in Natural Farming:

Integrating animal husbandry with agriculture, also known as mixed farming or integrated farming, is an effective way to maximize the productivity and profitability of small and large-scale natural farming systems. Below is a comprehensive strategy to integrate animal husbandry with agriculture:

12.2.1 Utilize Crop Residues as Feed:

Residue Usage: Use crop residues such as straw, husk, and leaves as fodder for livestock. For instance, paddy straw can be fed to cattle, and maize stover can be fed to goats and sheep. **Processing:** Convert residues into silage or hay during surplus periods to ensure feed availability in lean seasons.

12.2.2 Use Manure as Organic Fertilizer:

Composting: Collect animal manure and compost it to create nutrient-rich organic fertilizers for crops. **Bio-Digestion:** Install a biogas plant to use manure for energy generation and use the slurry as liquid fertilizer.

12.2.3 Crop Rotation with Fodder Crops:

Grow fodder crops like alfalfa, maize, or Napier grass in rotation with primary crops to ensure a year-round supply of green fodder. This also improves soil fertility through nitrogen fixation (e.g., legumes).

12.2.4 Integrate Poultry or Small Ruminants:

Poultry: Rear poultry in fields or near farmhouses. Poultry manure enriches the soil, while they control pests naturally. **Small Ruminants:** Sheep and goats can graze on weeds or fallow lands, reducing maintenance costs.

12.2.5 Agroforestry Systems:

Plant fast-growing trees (e.g., Leucaena, Gliricidia) or fruit trees on field boundaries to provide shade, fodder, and additional income from fruits and timber.

12.2.6 Integrated Pest Management (IPM):

Use ducks in paddy fields to control pests like insects and snails while reducing the need for chemical pesticides.

12.2.7 Dairy Integration:

Maintain dairy animals such as cows or buffaloes for milk production, which can be sold for direct income. Establish local milk processing units for value-added products like ghee, yogurt, or paneer.

12.2.8 Aquaculture and Livestock:

Combine fish farming with paddy cultivation (e.g., rice-fish farming) or use animal waste to fertilize ponds for aquaculture.

12.2.9 Adopt Zero-Waste Farming:

Ensure every output from one subsystem becomes an input for another, reducing costs and increasing sustainability. Example: Use poultry droppings in fish ponds or as fertilizer.

12.2.10 Technology and Training:

Use modern tools like feed formulation software and sensors to monitor animal health and productivity. Train farmers in breeding, feeding, and disease management practices.

12.2.11 Market Linkages:

Form cooperatives or farmer producer organizations (FPOs) to collectively sell produce, livestock, milk, and value-added products for better prices.

12.2.12 Government Schemes and Subsidies:

Avail schemes such as Rashtriya Krishi Vikas Yojana, National Livestock Mission, or NABARD loans for integrated farming initiatives.

CHAPTER 13

MIXED AQUACULTURE

13.1 Objectives of Integrating Aquaculture with Natural Farming:

Integrating aquaculture with natural farming agribusiness creates a synergistic relationship between water, fish, and crops, enhancing productivity, sustainability, and profitability. Here are the key objectives:

13.1.1 Maximizing Resource Utilization:

Utilize water resources efficiently by combining fish farming with crop cultivation. Increases the productivity of available land and water resources.

13.1.2 Nutrient Recycling:

Use nutrient-rich water from fish ponds to irrigate and fertilize crops. Reduces dependency on synthetic fertilizers and enhances soil fertility naturally.

13.1.3 Diversified Income Streams:

Generate additional revenue through the sale of fish, aquatic plants, and crops. Reduces financial risks and improves overall farm profitability.

13.1.4 Waste Management and Zero Waste:

Recycle agricultural residues (e.g., crop stalks, weeds) as fish feed and use fish waste as natural fertilizers. Creates a circular economy and minimizes environmental pollution.

13.1.5 Enhancing Farm Biodiversity:

Promote a diverse ecosystem by integrating aquatic and terrestrial farming components. Increases ecological balance and resilience to pests and diseases.

13.1.6 Improved Water Management:

Use aquaculture ponds as water reservoirs for irrigation and rainwater harvesting. Ensures efficient water usage and reduces water scarcity issues.

13.1.7 Sustainable Pest and Weed Control:

Use fish in paddy fields (e.g., rice-fish farming) to control pests and weeds naturally. Reduces reliance on chemical pesticides and herbicides.

13.1.8 Climate Resilience:

Build a resilient farming system that buffers against climate risks by diversifying production. Minimizes losses during extreme weather events.

13.1.9 Food Security:

Provide a source of high-quality protein (fish) alongside staple crops. Ensures nutritional security for farming households and local communities.

13.1.10. Employment Generation:

Create rural employment opportunities in fish farming, pond management, and crop production. Improves livelihoods and reduces rural-to-urban migration.

13.1.11. Low Input Costs:

Reduce farm input costs by leveraging natural synergies (e.g., fish feed from farm residues, organic fertilizers from fish waste). Makes the system more economical and self-sustaining.

13.1.12. Promotion of Organic Practices:

Foster eco-friendly farming by integrating aquaculture with natural farming methods. Attracts premium markets for organic and sustainably produced goods.

13.1.13. Adaptation of Local Ecosystems:

Enhance local ecosystems by introducing fish species and aquatic plants suited to the region. Promotes the conservation of indigenous biodiversity.

13.1.14. Value Addition and Market Access:

Develop value-added products like dried fish, fish oil, or organic crops irrigated with nutrient-rich water. Increases marketability and revenue potential for the farm.

13.1.15. Educational and Demonstration Potential:

Serve as a model for other farmers to adopt sustainable integrated farming systems. Facilitates knowledge-sharing and widespread adoption of natural farming principles.

13.2 Strategies for integrating Aquaculture with natural farming:

Integrating aquaculture with natural farming involves creating a mutually beneficial system where fish farming complements crop production and vice versa. Here are the best strategies to maximize profits while maintaining ecological sustainability:

13.2.1 Rice-Fish Farming:

Combine paddy cultivation with fish farming by releasing fish into the flooded rice fields. Fish control pests and weeds, reducing the need for chemicals. Nutrient cycling between fish waste and crops enhances soil fertility. Increases farm income through dual outputs (rice and fish). Recommended Fish: Tilapia, catfish, carp, or indigenous species like murrel.

13.2.2 Aquaponics Systems:

Use nutrient-rich water from fish tanks to grow vegetables, herbs, or leafy greens in a soilless setup. Fish provide natural nutrients for plant growth, while plants filter and purify the water for fish. Produces high-value crops like lettuce, basil, or strawberries alongside fish. Maximizes resource efficiency in small spaces.

13.2.3 Pond Irrigation for Crops:

Use fish ponds as reservoirs for irrigating crops, where nutrient-enriched water acts as a liquid fertilizer. Reduces fertilizer costs. Enhances crop yield and quality through organic inputs.

13.2.4 Duck-Cum-Fish Farming:

Rear ducks alongside fish in a pond system. Ducks naturally fertilize the pond and control weeds and insects. Ducks provide eggs and meat, adding an income stream. Fish benefit from the organic waste, promoting faster growth.

13.2.5 Livestock-Fish Integration:

Locate livestock sheds near fish ponds, allowing runoff from sheds to feed the fish indirectly. Poultry or goat droppings act as natural fish feed or fertilizer for aquatic plants. Generates income from livestock products like eggs, milk, or meat.

13.2.6 Composting Aquatic Waste:

Use fish scales, bones, and other waste to create high-quality compost for crops. Adds value to waste materials. Provides an organic alternative to chemical fertilizers.

13.2.7 Polyculture in Fish Ponds:

Cultivate multiple fish species in the same pond that occupy different ecological niches (e.g., surface, middle, and bottom feeders). Increases pond productivity. Reduces competition for food among fish species.

13.2.8 Floating Gardens:

Build floating rafts on fish ponds to grow vegetables like tomatoes, spinach, or cucumbers. Utilizes unused pond surface. Provides an additional income stream.

13.2.9 Integrated Pest and Weed Control:

Use fish to control weeds and pests in aquatic systems, reducing labor and pesticide costs. Saves on crop maintenance costs. Promotes a chemical-free farming environment.

13.2.10. Cultivating High-Value Fish Species:

Focus on cultivating high-demand fish species like prawns, trout, or ornamental fish for niche markets. Higher market prices increase profit margins. Diversifies income streams.

13.2.11. Biogas from Aquatic Waste:

Install a biogas plant to process fish waste and other organic farm residues into energy. Provides renewable energy for farm operations. Residual slurry can be used as organic fertilizer.

13.2.12. Value-Added Products:

Process fish into fillets, dried fish, or pickled products and sell crops as organic or premium-quality produce. Increases profitability through branding and higher market value.

13.2.13. Training and Certification:

Obtain organic certification for aquaculture and crop products and market them as eco-friendly. Attracts premium prices in niche organic markets.

13.2.14. Diversify with Agroforestry:

Combine aquaculture with agroforestry by planting fruit or fodder trees around ponds. Provides shade for ponds, reducing water evaporation. Generates additional income from fruits and timber.

13.2.15. Efficient Marketing and Value Chains:

Develop direct-to-consumer sales models or partner with Farmer Producer Organizations (FPOs) for better market access. Reduces dependency on middlemen, ensuring higher profits.

CHAPTER 14

FOOD PROCESSING

Food processing refers to the transformation of raw agricultural products into consumable food through various physical and chemical methods. This includes processes such as washing, chopping, cooking, canning, fermenting, and packaging and traditional methods, like drying and pickling.

14.1 Objectives of Food Processing:

The objectives of food processing are diverse, aiming to improve the quality, safety, availability, and value of food products. Here are the key objectives:

14.1.1 Preservation:

Extend Shelf Life: Reduce spoilage caused by microorganisms, enzymes, and oxidation. **Reduce Seasonal Dependency**: Ensure availability of food products year-round, irrespective of seasonal variations. **Maintain Nutritional Value**: Retain essential nutrients during storage and distribution.

14.1.2 Enhance Food Safety:

Eliminate Contaminants: Remove or destroy harmful bacteria, viruses, and toxins through techniques like pasteurization or sterilization. **Meet Regulatory Standards**: Comply with food safety laws and certifications.

14.1.3 Improve Food Quality:

Enhance Taste and Appearance: Use processing techniques to improve flavor, texture, color, and overall appeal. **Consistency**: Maintain uniform quality in large-scale production.

14.1.4 Add Value:

Transform Raw Materials: Convert raw agricultural products into consumer-friendly goods (e.g., wheat to flour, milk to cheese). **Create New Products**: Develop innovative food products to meet changing consumer demands.

14.1.5 Facilitate Distribution and Storage:

Ease of Transportation: Processed foods are often lighter, more compact, or less perishable, making them easier to transport. **Convenient Packaging**: Enhance storage and handling with packaging that protects food and extends shelf life.

14.1.6 Minimize Food Waste:

Utilize By-Products: Convert agricultural and processing by-products into usable forms (e.g., animal feed, biofuel). **Preserve Excess Produce**: Process surplus food to prevent wastage during harvest gluts.

14.1.7 Cater to Consumer Convenience:

Ready-to-Eat Options: Provide pre-cooked or easy-to-prepare food products to save consumers time and effort. **Portion Control**: Offer appropriately portioned products for individuals and families.

14.1.8 Support Economic Growth:

Job Creation: Create employment opportunities in farming, processing, packaging, and distribution. **Promote Exports**: Add value to raw materials, making them suitable for international markets.

14.1.9 Enhance Nutrition:

Fortification: Enrich food with additional nutrients like vitamins and minerals. **Diet-Specific Products**: Develop products catering to specific dietary needs (e.g., gluten-free, low-sodium).

14.1.10. Ensure Sustainability:

Eco-Friendly Practices: Implement sustainable methods to reduce environmental impact, such as energy-efficient processing or biodegradable packaging. **Reduce Carbon Footprint**: Optimize logistics and production to minimize resource use.

14.2 Methods of Food Processing:

Food processing involves various methods to transform raw ingredients into consumable or preservable forms. These methods can be broadly categorized into physical, chemical, thermal, biological, and advanced techniques. Below are the common methods of food processing:

14.2.1 Thermal Processing:

Pasteurization: Heating food to a specific temperature to kill harmful microorganisms (e.g., milk, juices). **Sterilization**: High-temperature treatment to destroy all microorganisms (e.g., canned foods). **Blanching**: Brief boiling followed by cooling to inactivate enzymes and retain color and texture (e.g., vegetables). **Drying/ Dehydration**: Removing water content to prevent spoilage (e.g., dried fruits, powdered milk). **Baking and Roasting**: Applying heat to achieve texture and flavor (e.g., bread, nuts).

14.2.2 Mechanical Processing:

Grinding and Milling: Reducing food into smaller particles or powders (e.g., wheat to flour). **Cutting and Chopping**: Reducing size for convenience and consistency (e.g., diced vegetables). **Emulsification**: Mixing two immiscible liquids (e.g., oil and water in mayonnaise). **Extrusion**: Forcing food through a shaped die to create specific forms (e.g., pasta, snack foods).

14.2.3 Preservation Techniques:

Freezing: Slowing microbial activity by lowering temperature (e.g., frozen vegetables, meat). **Canning:** Sealing food in airtight containers and heating to kill microorganisms. **Pickling:** Preserving food in acidic solutions like vinegar (e.g., pickles). **Salting and Curing:** Using salt to draw moisture out and inhibit bacterial growth (e.g., cured meats, fish). **Smoking:** Exposing food to smoke for flavor and preservation (e.g., smoked salmon).

14.2.4 Fermentation:

Using microorganisms (yeasts, bacteria) to convert sugars into alcohol, acids, or gases, enhancing flavor and shelf life (e.g., yogurt, kimchi, beer).

14.2.5 Chemical Processing:

Addition of Preservatives: Using chemicals like sodium benzoate to inhibit spoilage. **Fortification:** Adding nutrients to improve nutritional value (e.g., fortified cereals). **Acidification:** Lowering pH with acids like citric acid for preservation and flavor.

14.2.6 Advanced Processing Techniques:

High-Pressure Processing (HPP): Using high pressure to destroy pathogens without heat (e.g., cold-pressed juices). **Irradiation:** Using ionizing radiation to kill bacteria and pests and extend shelf life (e.g., spices, meat). **Ultrasonic Processing:** Using sound waves to mix, emulsify, or extract components (e.g., juices, oils). **Microwave Heating:** Quick, even heating using electromagnetic waves (e.g., ready-to-eat meals).

14.2.7 Bioprocessing:

Enzymatic Processing: Using enzymes to alter food properties (e.g., tenderizing meat, brewing beer). **Probiotics Incorporation:**

Adding beneficial bacteria to food for health benefits (e.g., probiotic yogurt).

14.2.8 Packaging Techniques:

Vacuum Packaging: Removing air to prevent oxidation and spoilage. **Modified Atmosphere Packaging (MAP):** Adjusting the gas composition inside the package to extend shelf life.

14.2.9 Mixing and Homogenization:

Homogenization: Breaking down fat molecules to create a uniform texture (e.g., milk). **Mixing/Blending:** Combining ingredients for consistency (e.g., cake batter, spice mixes).

14.2.10. Innovative Methods:

Freeze-Drying: Removing moisture at low temperatures to preserve nutrients and flavor (e.g., instant coffee). **Encapsulation:** Coating ingredients to protect them until consumption (e.g., vitamins in cereals).

CHAPTER 15

MARKETING MANAGEMENT IN NATURAL FARMING

15.1 Theories on Agricultural Marketing:

Theories on agricultural marketing provide insights into how agricultural products are distributed, sold, and valued in different market systems. These theories address the dynamics between producers, consumers, intermediaries, and external factors influencing market behavior. Below are the key theories:

15.1.1 Cobweb Theory:

Explains the cyclical fluctuations in prices and supply of agricultural products due to delays in production adjustments. Agricultural production has a time lag (e.g., crops take months to grow). Farmers base planting decisions on current prices. High prices in one season lead to overproduction in the next, causing a price drop. Low prices then result in underproduction, causing a price spike in subsequent seasons. Implication: Price and supply oscillations can lead to unstable agricultural markets.

15.1.2 Law of Comparative Advantage (David Ricardo):

Suggests that countries or regions should specialize in producing agricultural goods they can produce most efficiently and trade for other goods. Encourages regional specialization to maximize resource use.

15.1.3 Market Structure Theory:

Examines the organization of agricultural markets based on the number of buyers and sellers, product differentiation, and barriers to entry. Types of Market Structures: **Perfect Competition:** Many producers selling identical products (common in staple crop markets). **Monopolistic Competition:** Products are differentiated by brand or quality (e.g., organic vs. conventional produce). **Monopoly/ Oligopoly:** Few buyers (e.g., large supermarket chains) dominate the market. Power imbalances can lead to price manipulation and exploitation of farmers.

15.1.4 Institutional Approach:

Focuses on the role of institutions (e.g., cooperatives, marketing boards, policies) in shaping agricultural markets. Institutions provide support for price stabilization, quality control, and market access. Examples: Minimum Support Price (MSP) in India, marketing cooperatives for small-scale farmers.

15.1.5 Supply Chain and Value Chain Theories:

Analyze the flow of agricultural products from producers to consumers, adding value at each stage. **Supply Chain:** Emphasizes logistics, storage, and transportation. **Value Chain:** Focuses on value addition through processing, branding, and packaging. Efficient supply chains reduce post-harvest losses, while strong value chains increase farmers' incomes.

15.1.6 Risk and Uncertainty Theories:

Address the risks inherent in agricultural marketing due to factors like weather, pests, market volatility, and policy changes. **Portfolio Theory:** Farmers diversify crops or markets to minimize risks. **Expected Utility Theory:** Farmers make marketing decisions based on expected outcomes under uncertain conditions. Managing risks requires access to market information, insurance, and storage facilities.

15.1.7 Price Spread Theory:

Explains the difference between the price paid by consumers and the price received by farmers. Components: Costs of transportation, storage, processing, and marketing. Margins taken by intermediaries (e.g., wholesalers, retailers). Large price spreads can lead to farmer discontent and inefficiencies in the market.

15.1.8 Market Information Theory:

Highlights the role of information in agricultural markets. Accurate and timely information on prices, demand, and supply helps farmers make better marketing decisions. Information asymmetry can lead to exploitation by intermediaries. Use of ICT tools like mobile apps and online platforms to provide market updates.

15.1.9 Dual Economy Theory (W. Arthur Lewis):

Explains the coexistence of traditional (subsistence farming) and modern (commercial farming) sectors in agricultural economies. Transition from traditional to modern markets depends on investments in infrastructure, education, and technology.

15.1.10. Behavioral Theories:

Focus on the decision-making behavior of farmers and consumers in agricultural markets. **Prospect Theory**: Farmers weigh potential losses more heavily than gains when making marketing decisions. **Consumer Behavior Models**: Explain how consumers choose between conventional and organic/agroecological products based on price, quality, and brand. Behavioral insights can guide marketing strategies and policies.

15.1.11. New Institutional Economics:

Explores how transaction costs, property rights, and governance structures influence agricultural marketing. High transaction costs (e.g., transportation, negotiations) reduce farmers' net income. Contracts and cooperatives can lower transaction costs. Institutional reforms are essential for improving market efficiency.

15.1.12. *Theory of Agricultural Market Integration:*

Analyzes the interconnectedness of local, national, and international markets. Integrated markets have uniform prices after adjusting for transport costs. Fragmented markets show price disparities due to poor infrastructure or policies. Enhancing market integration improves price stability and farmer incomes.

15.2 Objectives of Marketing Management in Natural Farming:

Marketing management in natural farming focuses on promoting sustainable, eco-friendly agricultural practices while ensuring profitability for farmers and satisfaction for consumers. The objectives of marketing management in this sector aim to align environmental, social, and economic goals. Below are the key objectives:

15.2.1 *Promote Sustainability:*

Encourage the adoption and marketing of products that are grown using natural, chemical-free methods. Reduce the ecological footprint of agricultural marketing by using sustainable packaging, local sourcing, and low-carbon transportation methods.

15.2.2 *Enhance Market Access for Farmers:*

Provide natural farmers access to domestic and international markets through direct marketing, online platforms, and farmer-consumer networks. Reduce dependency on intermediaries to ensure fair pricing and better income for farmers.

15.2.3 Build Consumer Awareness:

Educate consumers about the benefits of natural farming products, including health, environmental, and ethical advantages. Use branding and labeling (e.g., "chemical-free," "organic," or "natural") to highlight product uniqueness.

15.2.4 Ensure Quality Assurance:

Maintain high standards of product quality to build trust among consumers. Implement certification systems (e.g., organic or natural farming certifications) to authenticate product claims.

15.2.5 Foster Direct Marketing Models:

Develop systems like farmer markets, community-supported agriculture (CSA), and farm-to-table programs to connect farmers directly with consumers. Minimize marketing costs and enhance farmer-consumer relationships.

15.2.6 Price Stability:

Protect farmers from price volatility by introducing mechanisms like minimum support prices, forward contracts, or group marketing. Offer consumers competitive pricing for premium natural farming products.

15.2.7 Promote Value Addition:

Encourage farmers to process, package, and brand their products (e.g., natural oils, herbal teas, or organic jams) to enhance market value. Support the creation of small-scale agro-industries to add value to raw produce.

15.2.8 Support Small and Marginal Farmers:

Develop inclusive marketing strategies tailored for small-scale natural farmers who may lack resources for large-scale operations.

Organize farmer cooperatives or producer groups to enable collective marketing and better bargaining power.

15.2.9 Facilitate Export Opportunities:

Tap into growing global demand for natural and organic products by promoting exports. Comply with international standards and certifications to access high-value markets.

15.2.10. Reduce Post-Harvest Losses:

Invest in cold storage, transportation, and processing facilities to minimize wastage. Improve supply chain management to ensure fresh and high-quality products reach consumers.

15.2.11. Encourage Digital and E-Marketing:

Use digital platforms (e.g., e-commerce websites, mobile apps) to market natural farming products directly to consumers. Promote online branding and social media marketing to reach a wider audience.

15.2.12. Focus on Ethical and Fair Trade Practices:

Ensure ethical sourcing and fair prices for farmers. Promote transparency in the supply chain to build consumer trust.

15.2.13. Support Community Development:

Use marketing revenues to invest in community welfare, such as better infrastructure, education, and healthcare for rural areas. Encourage farmers to adopt natural farming as a sustainable livelihood option.

15.2.14. Promote Local and Seasonal Produce:

Advocate for the consumption of locally grown, seasonal produce to reduce transportation costs and carbon footprint. Develop marketing campaigns emphasizing the freshness and health benefits of such products.

15.2.15. Align with Government Policies:

Leverage government schemes and subsidies for natural farming to support marketing efforts. Collaborate with government agencies for promotional campaigns, market access, and financial support.

15.2.16. Build Brand Loyalty:

Develop a strong brand identity for natural farming products, focusing on trust, sustainability, and health benefits. Engage consumers in long-term relationships through loyalty programs, subscriptions, or personalized services.

15.2.17. Drive Innovation in Marketing:

Experiment with innovative marketing strategies like storytelling, eco-labels, and experiential marketing (e.g., farm visits or virtual tours). Use blockchain technology for transparency in sourcing and production processes.

15.2.18. Integrate Sustainable Packaging:

Use biodegradable, recyclable, or reusable materials for packaging to align with the principles of natural farming. Highlight eco-friendly packaging as part of the product's value proposition.

15.2.19. Develop Cooperative Marketing Strategies:

Encourage the formation of natural farming cooperatives to pool resources and achieve economies of scale. Use collective branding to enhance visibility and reach for small-scale farmers.

15.2.20. Ensure Long-Term Profitability:

Develop marketing strategies that balance immediate profitability with long-term sustainability. Focus on building resilient markets for natural farming products to withstand economic and environmental challenges.

15.3 Methods of Marketing Management:

Marketing strategies for a natural farming agribusiness should focus on emphasizing sustainability, health benefits, and the unique value of the products. Here's a detailed plan:

Brand Identity:

Mission Statement: Farmers should create their own brands, highlighting their commitment to natural, chemical-free farming and its role in preserving the environment.

Brand Storytelling: Farmers should share stories about their farming journey, practices, and the positive impact on nature. **Visual Identity:** Farmers should use eco-friendly branding elements, like earthy tones and nature-inspired designs, for logos, packaging, and advertisements.

Audience Education:

Workshops & Seminars: Farmers should host events for farmers, gardeners, and health-conscious individuals to learn about natural farming methods and benefits.

Content Marketing:

Blog: Educated natural farmers should start writing blogs and articles on natural farming techniques. **Digital Content:** They should share educational videos, infographics, and farm-to-table stories. **Book:** They can offer books or booklets on Natural Farming as lead magnets or giveaways to attract customers.

Leveraging Digital Marketing:

Website and E-Commerce: Farmers should build an attractive website where customers can learn about their products and buy them directly. These websites should include customer reviews, certifications, and sustainability metrics to build trust. **Social Media Platforms:** Farmers should actively use social media platforms, such as, Instagram (visual content), Facebook (community), LinkedIn (B2B networking). **Content Ideas:** Farmers should show the natural farming process (from planting to harvesting) in their channels, post testimonials from satisfied customers and run contests for eco-conscious consumers.

Search Engine Optimization (SEO): Farmers should use SEO to rank for keywords like "natural farming products," "organic fertilizers," and "eco-friendly farming solutions."

Offer Products and Services Tailored to the Market:

Consumer-Focused Products: Farmers should develop their own unique organic produce, natural fertilizers, eco-friendly pesticides, and farming kits. **Farmer-Focused Solutions:** Farmers should start natural farming consultancy services, training programs and workshops. **Value-Added Products:** Farmers should produce processed goods like organic jams, oils, or herbal teas and ready-to-use natural farming inputs.

Certifications and Accreditations:

Farmers should obtain certifications like: Organic certification, Good Agricultural Practices (GAP) or Eco-friendly packaging credentials. They should use these on packaging, websites, and marketing materials to build trust.

Build Strategic Partnerships:

Collaborate Locally: Farmers should partner with organic markets, restaurants, and health stores and they should work with NGOs promoting sustainable farming practices. **Institutional Partnerships:** Farmers should supply produce to schools, hospitals, and corporate cafeterias promoting healthy eating.

Use Direct-to-Consumer (D2C) Channels:

Subscription Boxes: Farmers should explore direct-to-home channels to deliver fresh organic produce to consumers' homes on a subscription basis. **Farm Tours:** Farmers should invite consumers to visit their farms and experience the process firsthand.

Participate in Events and Trade Shows:

Farmers should attend agriculture expos, natural product fairs, and sustainability forums. They should use these platforms to demonstrate your products, build connections, and gain customers.

Utilizing Influencer and Word-of-Mouth Marketing:

Farmers should collaborate with eco-conscious influencers, chefs, and food bloggers. Offer referral discounts or incentives to customers who recommend their products.

Customer-Centric Strategies:

Farmers should offer loyalty programs for repeat customers, use surveys to gather feedback and improve and should provide excellent customer service to build long-term relationships.

CHAPTER 16

ECO-AGRITOURISM WITH NATURAL FARMING

Eco-agritourism, also known as agri-ecotourism, combines agricultural experiences with eco-friendly practices, allowing visitors to engage with sustainable farming while enjoying nature. This form of tourism promotes environmental conservation and supports local communities by showcasing farming methods that protect ecosystems and biodiversity. This approach not only provides educational and recreational opportunities but also promotes sustainable agricultural practices and supports local economies (Wiranatha AS et al. 2024).

16.1 Objectives of Eco-Agritourism:

The objectives of eco-agritourism are purely economic, to create multiple sources of income for the farmer. Here are the main objectives:

16.1.1 Transition from Product Manufacturing to development of Complete Business Ecosystem:

The objective of eco-agritourism is to explore all options to multiply the sources of income for the farmers, by integrating all verticals of natural farming agribusiness.

16.1.2 Maximise Tourist Visit, Profit per Tourist and Repeat Booking:

The ultimate goal of eco-agritourism is to maximise tourist visit throughout the year, to maximise the revenue per tourist and to maximise repeat visit and repeat buying by the visitors.

16.1.3 Monetize Natural Farming Agroecosystem:

A well managed natural farming agroecosystem often looks like a Botanical Garden, a Zoological Garden, a Biodiversity Park (Eco-Park or Nature Park), an Aqua Park, a Permanent Flower Show and a Supermarket of Agricultural Goods. The farmer can Monetize cultural ecosystem services of the natural farming agroecosystems. Generate additional income for farmers by diversifying their revenue streams. Support rural economies by creating jobs in tourism, hospitality, and related sectors. Promote local products and crafts, enhancing market access for rural producers. Encourage investment in rural infrastructure and services.

16.1.4 Environmental Conservation::

Promote sustainable agricultural practices that protect soil, water, and biodiversity. Reduce the carbon footprint of tourism by focusing on eco-friendly operations. Encourage organic farming and natural farming techniques to restore ecosystems. Foster appreciation for nature and conservation among visitors.

16.1.5 Education and Awareness::

Educate visitors about sustainable farming methods, natural ecosystems, and the importance of biodiversity. Provide hands-on experiences such as planting, harvesting, or composting to foster understanding of eco-friendly agriculture. Raise awareness about food security, healthy eating, and the farm-to-table concept. Promote knowledge-sharing between farmers, researchers, and tourists.

16.1.6 Cultural Preservation:

Highlight the traditions, customs, and heritage of rural communities. Preserve traditional farming practices, tools, and knowledge systems. Promote local cuisines and festivals as part of the tourism experience. Encourage the continuation of indigenous farming techniques and lifestyles.

16.1.7 Strengthening Farmer-Community Relations:

Foster stronger connections between urban and rural populations. Create platforms for dialogue about agricultural challenges and sustainable solutions. Involve local communities in decision-making and profit-sharing.

16.1.8 Promoting Wellness and Health:

Offer visitors opportunities to experience fresh, organic food directly from farms. Provide tranquil, nature-based environments for relaxation, mental rejuvenation, and physical health. Highlight the role of agriculture in producing nutrient-dense foods that boost health and immunity.

16.1.9 Advocacy for Climate-Resilient Agriculture:

Demonstrate farming techniques that adapt to and mitigate climate change impacts. Encourage practices like crop rotation, agroforestry, and water conservation. Showcase renewable energy solutions such as solar-powered irrigation and biogas.

16.1.10 Enhancing Biodiversity:

Develop agroecosystems that support pollinators, birds, and other wildlife. Integrate farming with natural habitats to create a harmonious environment. Promote the cultivation of diverse crops and traditional seeds to enrich genetic biodiversity.

16.1.11 Recreational and Aesthetic Value:

Provide tourists with an escape from urban life into serene rural settings. Offer activities like bird-watching, trekking, fishing, or camping in eco-friendly farm environments. Enhance the aesthetic appeal of rural areas through agro-landscaping and natural beauty.

16.1.12. Sustainable Tourism Development:

Promote responsible tourism that minimizes environmental impact and respects local cultures. Set an example of a tourism model that balances economic benefits with ecological integrity. Encourage eco-conscious behavior among tourists, such as waste reduction and respect for nature.

16.2 Strategies for Profit Maximisation from Eco-Agritourism:

Eco-agritourism combines sustainable agriculture with tourism, offering farmers a lucrative avenue to maximize profits while promoting environmental conservation and rural development. Here are the best strategies for profit maximization in eco-agritourism:

16.2.1 Develop Unique, Authentic Experiences:

Farm Tours: Offer guided tours showcasing organic farming techniques, natural composting, and permaculture systems. **Hands-On Activities**: Allow visitors to participate in activities like planting, harvesting, milking, or fruit picking. **Cultural Integration**: Highlight local traditions, cuisine, and festivals.

16.2.2 Create Value-Added Products:

Farm-Fresh Produce: Sell organic fruits, vegetables, and herbs directly to tourists. **Processed Goods**: Offer value-added products like jams, pickles, organic teas, and oils. **Handicrafts**: Collaborate with local artisans to sell handmade goods.

16.2.3 Focus on Eco-Friendly Infrastructure:

Sustainable Accommodations: Build eco-lodges, cottages, or treehouses using natural materials. **Renewable Energy:** Use solar panels, biogas, or wind energy to power the facilities. **Water Management:** Install rainwater harvesting systems and promote water conservation.

16.2.4 Leverage Farm-to-Table Dining:

Farm Restaurants: Serve meals made from fresh, organic produce grown on-site. **Cooking Workshops:** Teach tourists how to prepare traditional dishes using farm produce. **Health and Wellness Focus:** Highlight the nutritional and environmental benefits of organic food.

16.2.5 Offer Educational and Recreational Programs:

Workshops and Seminars: Organize sessions on sustainable farming, agroforestry, or beekeeping. **Children's Activities:** Create kid-friendly programs like petting zoos, tractor rides, or nature trails. **Nature-Based Experiences:** Introduce bird watching, star gazing, or wildlife spotting in nearby areas.

16.2.6 Diversify Revenue Streams:

Membership Programs: Offer seasonal passes for repeat visits or discounts on produce. **Subscription Services:** Provide weekly or monthly deliveries of organic produce to urban customers. **Corporate Tie-Ups:** Host team-building retreats or eco-conferences.

16.2.7 Marketing and Branding:

Storytelling: Share your journey, values, and eco-agritourism mission through social media, blogs, and videos.

Local Collaborations: Partner with nearby tourist attractions, homestays, or eco-tourism operators. **Digital Presence**: Create a user-friendly website with booking options and promote via platforms like Instagram, Facebook, and TripAdvisor.

16.2.8 Promote Wellness and Relaxation:

Yoga and Meditation Retreats: Host wellness sessions in serene, natural surroundings. **Healing Gardens**: Design areas for relaxation with medicinal and aromatic plants. **Herbal Therapy**: Offer herbal teas, oils, and products for rejuvenation.

16.2.9 Collaborate with Local Communities:

Employment Opportunities: Hire locals to support farming, guiding, or hospitality operations. **Cultural Performances**: Invite local artists for music, dance, or storytelling sessions. **Craft Markets**: Set up spaces for local artisans to sell their products.

16.2.10. Prioritize Guest Satisfaction:

Personalized Services: Cater to individual preferences, dietary needs, and accessibility requirements. **Feedback Mechanism**: Act on visitor suggestions to improve offerings. **Loyalty Programs**: Reward repeat customers with discounts or exclusive perks.

Key Metrics for Success:

Visitor Numbers: Track the volume and frequency of tourist visits and number of repeat visitors. **Revenue per Visitor**: Increase spending by offering bundled experiences (e.g., stay, meals, activities). **Sustainability Goals**: Ensure your practices align with eco-friendly principles to attract conscious travelers.

CHAPTER 17

ECOLOGICAL ECONOMICS OF NATURAL FARMING

Ecological economics is an interdisciplinary field that integrates ecology and economics to understand and manage the relationships between human economies and natural ecosystems. It challenges traditional economic theories by emphasizing sustainability, resource limits, and the interdependence of economic systems and the environment.

17.1 Theories of Ecological Economics:

Here are some key theories and concepts in ecological economics:

17.1.1 Steady-State Economy:

Proposed by Herman Daly, this theory advocates for an economic system with stable resource consumption and population levels.

Emphasizes maintaining ecological balance by limiting material throughput (resource extraction and waste). Contrasts with traditional growth-focused economics, arguing that continuous growth is unsustainable on a finite planet.

17.1.2 The Entropy Law and Economic Processes:

Introduced by Nicholas Georgescu-Roegen, this theory applies the second law of thermodynamics (entropy) to economics.

Argues that economic processes transform high-quality energy and materials into lower-quality, less usable forms, leading to irreversible resource depletion.

17.1.3 The Safe Operating Space for Humanity:

Popularized by the Planetary Boundaries Framework by Johan Rockström and colleagues. Identifies critical environmental thresholds (e.g., climate change, biodiversity loss) that humanity must stay within to ensure a stable Earth system. Encourages sustainable practices to prevent crossing these boundaries.

17.1.4 Weak vs. Strong Sustainability:

Weak Sustainability: Assumes that natural and human-made capital are substitutable, so resource depletion can be offset by technological advancements or artificial capital. **Strong Sustainability:** Argues that natural capital (ecosystems, biodiversity) is irreplaceable and must be preserved for long-term survival.

17.1.5 The Precautionary Principle:

Advocates for caution in economic activities that could harm the environment or human health, even if scientific evidence is incomplete. Encourages policies to prevent irreversible damage.

17.1.6 Degrowth:

A social and economic movement that advocates reducing production and consumption to achieve sustainability and improve well-being. Challenges the notion that economic growth is essential for societal progress.

17.1.7 Social Metabolism:

Analyzes the flow of energy and materials through economies, viewing them as systems embedded within the biosphere. Highlights the need to balance resource extraction and waste assimilation with ecological limits.

17.1.8 Co-evolution of Human and Natural Systems:

Suggests that human economies and natural ecosystems evolve together, influencing and shaping each other. Recognizes the dynamic and interconnected nature of ecological and economic systems.

17.1.9 Ecological Footprint:

Developed by Mathis Wackernagel and William Rees, this concept measures the amount of biologically productive land and water required to sustain a population's consumption and absorb its waste. Highlights the overshoot of Earth's carrying capacity.

17.1.10 The Doughnut Economics Model:

Proposed by Kate Raworth, this framework visualizes a safe and just space for humanity. Combines social foundation (minimum standards for well-being) with ecological ceiling (planetary boundaries) to guide sustainable development.

17.1.11 The Resilience Theory:

Explores the capacity of ecosystems and economies to withstand shocks and adapt to changes without collapsing. Emphasizes maintaining diversity and flexibility in systems to ensure sustainability.

17.2 Application of Ecological Economics in natural farming:

Ecological economics integrates economic principles with ecological understanding, emphasizing sustainability, resource efficiency, and long-term value creation. When applied to natural farming, it can help maximize profits while preserving the ecosystem. Here's how:

17.2.1 *Optimize Resource Efficiency:*

Use resources efficiently to reduce costs and improve output without harming the environment. Implement rainwater harvesting, mulching, cover cropping, multilayer polycropping, mixed animal husbandry to maintain soil health and to maintain natural control of insect pests, diseases and weeds.

17.2.2 *Leverage Ecosystem Services:*

Capitalize on natural processes that support farming. Use pollinator biodiversity to improve yields in crops like fruits, vegetables, and oilseeds. Employ insect biodiversity (predators and parasitoids) to natural control pests, eliminating chemical pesticide expenses. Maintain soil biodiversity (earthworms, microorganisms) for nutrient cycling and better crop productivity.

17.2.3 *Diversify Revenue Streams:*

Reduce risk and enhance profitability through multiple income sources. Grow high-value, low-input crops like medicinal plants, herbs, or indigenous grains. Produce and sell value-added products (e.g., organic honey, herbal teas, pickles). Combine natural farming with eco-agritourism to attract visitors and generate additional revenue.

17.2.4 *Emphasize Circular Economy:*

Reuse and recycle farm outputs to minimize waste and maximize profits. Convert farm residues into mulch for soil enhancement. Use animal waste to produce biogas, reducing energy costs. Create zero-waste farming systems by linking livestock, crops, and waste management.

17.2.5 Enhance Market Access:

Capitalize on the growing demand for organic and sustainable products. Obtain organic certifications to access premium markets. Build direct-to-consumer sales channels (e.g., farmer markets, online stores). Partner with cooperatives or community-supported agriculture (CSA) programs to ensure consistent demand.

17.2.6 Adopt Ecosystem Valuation:

Account for the benefits provided by the ecosystem to determine the true value of natural farming. Showcase carbon sequestration benefits of natural farming to attract subsidies or carbon credits. Promote the farm as a biodiversity hotspot to secure government grants or eco-agritourism opportunities. Highlight health benefits of chemical-free produce to secure premium pricing.

17.2.7 Invest in Knowledge and Collaboration:

Share and learn sustainable practices to improve productivity and reduce costs. Collaborate with research institutions for training in advanced natural farming methods. Participate in farmer collectives to share resources, reduce costs, and improve market reach. Use digital tools for precision farming to optimize input use and reduce waste.

17.2.8 Promote Regenerative Practices:

Regenerate degraded lands for long-term profitability. Use agroforestry to restore soil health, improve water retention, and increase crop diversity. Grow nitrogen-fixing plants (e.g., legumes) to reduce fertilizer dependency. Implement contour farming and mulching to prevent soil erosion.

17.2.9 Incorporate Environmental Accounting:

Track and monetize environmental benefits of natural farming. Create a farm-level ecological balance sheet to highlight the positive impacts of natural farming. Use these metrics to attract eco-conscious investors, grants, or CSR funding. Quantify savings from reduced chemical inputs and energy use.

17.2.10 Build Community Support:

Foster community-based initiatives for mutual growth and cost-sharing. Share resources like machinery, seeds, and storage facilities through cooperative models. Host workshops or tours to educate others, earning extra income while building goodwill. Engage in collective bargaining for better prices on inputs and outputs.

Benefits of Ecological Economics in Natural Farming: (1) **Cost Reduction:** Eliminate dependency on external inputs like synthetic fertilizers and pesticides. (2) **Premium Pricing:** Higher demand and prices for organic, eco-friendly products. (3) **Sustainability:** Preserves the land's productive capacity for future generations. (4) **Market Differentiation:** Stronger appeal to eco-conscious consumers and investors.

CHAPTER 18

RESEARCH AND INNOVATION ON NATURAL FARMING

The concepts of research and innovation in natural farming (polyculture or biodiversity based agriculture) is a completely different paradigm from the concept of research and development conventional agriculture (monoculture or anti-biodiversity agriculture). Complete absence of tillage and external inputs in natural farming makes the present paradigm of research completely redundant. Due to this conceptual vacuum, research and innovation on natural farming remained a non-starter which is reflected in the complete absence of scientific literature on natural farming in the peer reviewed journals. A new framework and protocol of natural farming research has to be developed, to start research on natural farming. Here I fundamentally rethink and re-conceptualise the agricultural research paradigm, and develop a model framework of natural farming research, by analysing traditional agricultural knowledge of the indigenous people and modern protocol for ecological research.

18.1 Theories of Innovation:

18.1.1 4Ps of Innovation (Bessant and Tidd):

Categories of Innovation: **Product:** New goods or services (e.g., electric vehicles). **Process:** Better production methods (e.g., automation). **Position:** Repositioning products in the market (e.g., branding). **Paradigm:** Fundamental changes in business models (e.g., Airbnb).

18.1.2 Triple Helix Model:

Innovation arises from collaboration between academia, industry, and government.

18.1.3 Frugal Innovation (Jugaad Innovation):

Focuses on creating cost-effective solutions, especially in resource-constrained environments.

18.1.4 Business Model Innovation:

Innovating not just products but also how businesses deliver value. Examples: Subscription models (Netflix). Freemium models (Spotify).

18.1.5 Blue Ocean Strategy (W. Chan Kim and Renée Mauborgne):

Encourage companies to create new markets (blue oceans) rather than competing in saturated ones (red oceans).

18.1.6 Incremental Innovation:

Incremental innovation is a strategy focused on making small, gradual improvements to existing products, services, or processes. This approach emphasizes customer feedback and aims to enhance functionality, efficiency, and quality without the risks associated with radical changes.

18.1.7 Disruptive Innovation Theory (Clayton Christensen):

Innovations that initially target underserved or niche markets and eventually disrupt established industries. Examples: Personal computers disrupting mainframes, streaming services replacing DVDs.

18.2 Objectives of Research on Natural Farming:

The objective of conventional agricultural research is to maximise the use of commercial agricultural inputs and machines in agriculture, for profit maximisation of the input and machine manufacturing industries. In contrast, the objectives of natural farming research includes profit maximisation of the farmers, maximising public health, maximising soil and environment health, mitigation of climate change and biodiversity loss, and advancement of science. Here are some of the main objectives of research on natural farming:

18.2.1 Decentralisation and Democratisation of Agricultural Research:

Conventional system of agricultural research promotes centralisation and monopoly in agricultural research. Non-farmer scientists collect traditional plant varieties (landraces and wild progenitors) from farmers' fields without taking any consent from the farmers and without paying any royalty to the farmers for their intellectual property rights (IPR). Non-farmer scientists declare these plant genetic resources as common heritage of mankind and they maintain these genetic resources in various national and international ex-situ gene banks. They develop genetically modified (GM) or hybrid varieties from these genetic resources, by applying sophisticated methods of genetic engineering, for maximising the efficiency of these plant varieties in consuming chemical fertilisers, herbicides and pesticides. Hybrid (F1) varieties lose their hybrid vigour in the next (F2) generation, and this phenomenon helps rob the farmers rights to seed and seed sovereignty. Non-farmer scientists experiment with different combinations of synthetic chemical fertilisers, micronutrients, plant growth regulators, herbicides, pesticides, fungicides, antibiotics, irrigation systems, drones, and other machines, to maximise the profit of the seed, chemical and machine manufacturing industries, to help establish their monopoly.

In contrast, natural farming research promotes decentralised, independent agricultural research and innovation by individual farmers, for development of exclusive plant varieties and for development of exclusive cropping systems for individual farmers, for their own profit maximisation.

Development of Nature Based Solutions (NbS) for Agriculture:

Natural farming research conducted by farmers and scientists can develop modern nature based solutions (NbS) for sustainable agriculture. Nature based solutions are those solutions which uses natural resources and living organisms such as plants, animals and microorganisms (biology and ecology), instead of using chemicals (chemistry) or machines (physics or engineering), to solve the problems of mankind.

18.2.2 De Novo Domestication of Plants, Animals and Microorganisms:

Indigenous people started animal and plant domestication in the Neolithic Era, more than 10,000 years ago. They have domesticated thousands of plant species (exact number not known) and bred millions of unique plant varieties for the benefit of future generations. Most of these genetic resources have already been wiped out by the large-scale adoption of monoculture all over the world. This is known as genetic erosion. The traditional process of plant domestication remained inconclusive. Only a few plant species have been fully domesticated, some remained semi-domesticated and a large number of plants still remained as non-domesticated. Natural farming research should take it as a pririty to resume the process of traditional plant domestication, to develop novel plant varieties from lesser known indigenous fruits and vegetables.

18.2.3 Development of Exclusive Plant Varieties for Each Individual Farmer:

Conventional agricultural research develops plant varieties for all the farmers in a large geographic region (agro-climatic zone), to help the seed companies achieve economies of scale. In contrast, the objective of natural farming research is to develop unique, exclusive and exquisite plant varieties for each individual farmer, for profit maximisation of the farmers.

18.2.4 Rating and ranking of Plant Varieties and Animal Breeds:

Scientists should develop and standardise a system of rating and ranking of crop plant species and plant varieties of the world and this rating should be published in an open access, online database. This rating should be based on the nutritive value, medicinal value, economic value and ecological value of the plant varieties. This rating will help use and conservation of lesser known plant varieties, improvement of health of the consumers and profit maximisation of the farmer.

18.2.5 Plant and Animal Portfolio Optimisation:

The objective of modern natural farming research is to develop an artificial intelligence (AI) and machine learning (ML) based software for the farmers, to provide free personalised advice to each individual farmer to help them design, optimise, rebalance, and maintain their most appropriate portfolio of plant varieties and animal breeds, ultimately for their profit maximisation. Because, natural farming agroecosystems are highly complex ecosystems and farmers are in general less educated than others.

18.2.6 Development of Innovative Cropping Systems for Each Individual Farmer:

The objective of natural farming research is to develop unique and exclusive cropping systems for each individual farmer, because the location, soil, climate and market conditions of each individual farm is unique and completely different from the others.

18.2.7 Probiotics Innovation:

The objective of natural farming research is to empower each individual farmer with the knowledge and expertise, to help them develop their own unique crude microbial culture (probiotic formulations) from their available natural resources, for solving their day to day problems in crop production, crop protection and food processing.

18.2.8 Food Processing Innovation:

The objective of natural farming research is to empower and train each individual farmer to experiment with their own traditional methods of food processing and to develop (innovate) new processes and new products, for their profit maximisation.

18.2.9 Eco-Agritourism Innovation:

The objective of natural farming research is to develop innovative nature based solutions for maximising income of the farmers from eco-agritourism. Each natural farm should be unique and exclusive from others, to attract eco-agritourism. The decentralised system of natural farming research should empower each and every individual farmer to develop their own strategies for maximising eco-agritourism and for maximising their profit.

18.2.10. Development of Marketing Strategy for Each Individual Farmer:

The objective of natural farming research is to innovate exclusive nature based solutions for marketing of the natural farming products. The solutions might include market research, market segmentation, customer acquisition, customer retention, quality control, price discovery, and export.

18.2.11. Valuation of Natural Capital:

The objective of natural farming research is to take inventory and make a valuation of natural capital (natural resources) of the farm. This valuation of natural capital will help monitor the success and achievement of the natural farming agroecosystem.

18.2.12. Valuation of Ecosystem Services:

The final objective of natural farming research would be valuation of the ecosystem services of the farm. The ecosystem services include provisioning ecosystem services, regulating ecosystem services and cultural ecosystem services. This valuation will help assess the success of the natural farming agroecosystem.

18.2.13. Financial Accounting of Natural Farming Agribusiness:

The objective of natural farming research is to develop a new accounting standard for accounting the profit and loss accounts and balance sheet of the natural farming agribusiness. Because, a natural farming agroecosystem is a complex ecosystem like a supermarket with multiple products and functions.

18.2.14. Developing Hypotheses and Theories on Natural Farming:

The ultimate objective of modern natural farming research would be to re-evaluate the theories and hypotheses of ecology in respect of natural farming and to develop new hypotheses and theories on natural farming for the purpose of advancement of science. These new hypotheses and theories may finally evolve into new branches of agricultural science such as ecological agriculture, theoretical agriculture, mathematical agriculture, agricultural informatics, computational agriculture or quantum agriculture.

18.3 Methods of Natural Farming Research:

Scientific research on natural farming is still non-existent among the scientific communities as well as in reputed peer reviewed journals. The primary reason is the conceptual vacuum about the paradigm shift from input-based agriculture to biodiversity based agriculture. Here, a new framework of natural farming research and innovation was proposed, which was based on agroecosystem analysis of the traditional agricultural landscapes of the world.

18.3.1 Agroecosystem Analysis of Traditional Agricultural Landscapes for Initiating Research:

Conducting an agroecosystem analysis (AEA) of traditional agricultural landscapes provides a comprehensive understanding of their ecological, social, and economic dynamics, serving as a foundation for research initiatives. Below is an approach for initiating AEA: (1) **Define the Scope of Analysis: Purpose:** Understand specific objectives (e.g., promoting sustainability, biodiversity, or productivity). **Scale:** Determine the geographic scale (village, district, region). **Timeframe:** Consider both seasonal and long-term aspects of traditional practices. (2) **Collect Baseline Data:** (a) **Ecological Data: Crop Diversity:** Identify major and minor crops, cropping patterns, and intercropping practices. **Soil Health:** Analyze soil properties (texture, nutrients, organic matter). **Water Resources:** Map irrigation systems, sources, and water-use efficiency. **Biodiversity:** Assess flora and fauna, including pollinators and pest predators. (b) **Social and Cultural Data: Traditional Knowledge:** Document indigenous practices, rituals, and beliefs tied to farming. **Community Involvement:** Identify roles of farmers, women, and marginalized groups. **Decision-Making:** Study collective versus individual decision-making processes. (c) **Economic Data: Farm Economics:** Assess input-output costs, profitability, and market access. **Resource Allocation:** Study landholding patterns and labor use. **Value Chains:** Explore processing, storage, and sales practices. (3) **Identify Key Drivers and Constraints: Natural Drivers:** Rainfall patterns, climatic factors, and natural calamities. **Anthropogenic Drivers:** Policy changes, urbanization, and market demands. **Constraints:** Land degradation, water scarcity, or loss of traditional knowledge. (4) **Analyze Interactions and Relationships: Ecosystem Services:** Examine provisioning (food, fodder), regulating (pest control, water cycling), and cultural services. **Input–Output Flows:** Map nutrient cycles, energy use, and resource flows. **Resilience:** Study system adaptability to changes like climate variability. (5) **Synthesize Insights for Research Focus: Problem Identification:** Highlight specific issues needing attention (e.g., soil erosion, declining yields). **Research Opportunities:** Integration of modern technologies with traditional practices. Exploring climate-smart agricultural practices. Enhancing biodiversity for natural pest control. Improving water management. (6] **Develop Participatory Framework: Stakeholder Engagement:** Involve farmers, local institutions, and policymakers. **Capacity Building:** Train farmers in sustainable techniques and documentation. **Feedback Mechanisms:** Regularly incorporate farmers' feedback into the research process. (7) **Documentation and Dissemination:** Use geospatial tools (e.g., GIS) for mapping. Publish findings in accessible formats for policymakers and practitioners. Share results through community workshops and social media platforms.

18.3.2 Characterisation of Traditional Agricultural Knowledge for Initiating Research:

Characterizing Traditional Agricultural Knowledge (TAK) is crucial for preserving indigenous practices, understanding their scientific basis, and integrating them into research for sustainable agriculture. Below is a stepwise framework for characterizing TAK to initiate research: (1] **Define Objectives and Scope: Purpose:** Identify why TAK is being studied (e.g., sustainability, climate resilience, biodiversity enhancement). **Scope:** Determine focus areas such as crop production, pest management, water conservation, or agroforestry. **Scale:** Specify geographic (local, regional) and temporal (seasonal, historical) boundaries. (2) **Data Collection Methods:** (a) **Documentation of Practices: Crop Management:** Record traditional methods for planting, intercropping, rotation, and harvesting. **Soil Management:** Include practices like composting, mulching, and use of biofertilizers. **Pest and Disease Control:** Note the use of natural repellents, trap crops, and biological control. **Water Management:** Document techniques like rainwater harvesting, canal systems, and flood-based farming. (b) **Ethnographic Studies: Interviews and Focus Groups:** Engage with elders, farmers, and community leaders to gather oral histories and contextual knowledge. **Participant Observation:** Observe farming activities and festivals linked to agricultural cycles. (c) **Literature Review:** Study local historical records, manuscripts, and existing research on indigenous practices. [3] **Characterization and Categorization:** (a) **Scientific Validation:** Analyze practices scientifically to identify their ecological, agronomic, and economic benefits. Compare TAK with modern agricultural practices to find complementary applications. (b) **Classification: Thematic Categories:** Crop management, soil fertility, pest control, water conservation, etc. **Temporal Relevance:** Seasonal versus year-round practices. **Cultural Context:** Religious or ritual significance of certain practices. (c) **Geospatial Analysis:** Map regional variations of TAK to identify unique practices and their environmental correlations. [4] **Analyze Strengths and Limitations: Strengths:** Sustainability and resilience. Low-cost and locally available solutions. Biodiversity conservation. **Limitations:** Decline in knowledge transmission due to modernization. Challenges in scalability and commercial viability. (5) **Establish Linkages for Research: Agroecological Relevance:** Align TAK with climate-smart agriculture, soil health improvement, and biodiversity preservation. **Interdisciplinary Integration:** Incorporate knowledge into agronomy, ecology, and socio-economic studies. **Technological Augmentation:** Explore how traditional methods can be enhanced using modern tools (e.g., drones for monitoring traditional irrigation). (6) **Community Participation: Farmers as Collaborators:** Actively involve farmers in designing experiments and validating practices. **Capacity Building:** Train communities to document and refine their traditional knowledge systems. **Recognition and Incentives:** Acknowledge contributors through publications, certifications, or financial incentives. (7) **Documentation and Dissemination: Databases:** Create open-access repositories of TAK for researchers and practitioners. **Visual Media:** Use videos, infographics, and participatory documentaries to preserve knowledge. **Policy Advocacy:** Engage policymakers to incorporate TAK into sustainable agriculture policies. (8) **Research Opportunities Derived from TAK:** Study TAK's role in enhancing climate resilience. Develop biopesticides and biofertilizers from traditional formulations. Examine TAK's contribution to agroforestry and integrated farming systems. Promote TAK for organic certification and niche marketing.

18.3.3 Farmer-Scientist Collaboration (Co-Creation):

Farmer-scientist collaborations in research on Traditional Agricultural Knowledge (TAK) create a mutually beneficial framework to integrate indigenous practices with modern scientific approaches. These collaborations provide opportunities to advance sustainable agriculture, biodiversity conservation, and climate resilience. The farmers goal is to develop new plant varieties and to develop new cropping systems for their profit maximisation out of their natural farming. While scientists' goal will be to develop new hypotheses or theories on natural farming for advancement of science. Below are key opportunities for such partnerships: (1) **Participatory Research Projects: On-Farm Trials:** Scientists can conduct trials of TAK practices directly on farmers' fields to evaluate their effectiveness under real-world conditions. Examples: Testing traditional pest control methods, indigenous crop varieties, or soil amendments. **Co-Creation of Solutions:** Farmers and scientists jointly design experiments, blending TAK with modern tools (e.g., integrating traditional irrigation with drip systems). (2) **Documentation and Validation of TAK: Knowledge Documentation:** Farmers provide oral histories and detailed descriptions of traditional practices, which scientists record and analyze. **Scientific Validation:** Scientists evaluate TAK methods for scalability, cost-effectiveness, and ecological impact (e.g., biofertilizers or natural pest repellents). **Case Study Development:** Collaborations can generate case studies showcasing successful TAK practices. (3) **Development of Climate-Resilient Practices: Adaptation Strategies:** Farmers share traditional drought, flood, and pest-resilient practices. Scientists refine and optimize these methods using predictive models and technology. **Seed Banks:** Establish community seed banks for indigenous varieties with farmer input on their performance and adaptability. (4) **Technology Integration with TAK: Digital Platforms:** Develop mobile apps or online tools where farmers input TAK methods, and scientists provide real-time feedback. **Precision Agriculture:** Scientists use remote sensing and IoT to enhance TAK-based irrigation or pest management systems. (5) **Enhancing Soil and Biodiversity Conservation: Soil Health**

Programs: Combine farmers' composting and organic fertilization techniques with scientific soil testing to optimize fertility. **Biodiversity Studies:** Collaborate on projects to catalog and protect agro-biodiversity, including traditional crop varieties and associated ecosystems. (6) **Policy Advocacy and Knowledge Sharing: Community Engagement Workshops:** Farmers and scientists co-host events to share findings and promote TAK among broader farming communities. **Policy Development:** Jointly develop policy recommendations to incorporate TAK into sustainable agriculture frameworks. (7) **Value Addition and Market Linkages: Value-Added Products:** Collaborate to develop products (e.g., organic fertilizers, biopesticides) from TAK practices, enhancing farmers' income. **Organic Certification:** Scientists support farmers in obtaining certifications for traditional methods meeting organic or ecological standards. **Market Development:** Establish niche markets for traditional crops or TAK-based products. (8) **Training and Capacity Building: Farmer–Scientist Exchange Programs:** Organize visits to research institutions for farmers and field trips for scientists to traditional farms. **Workshops and Seminars:** Jointly conduct training on TAK, integrating modern scientific insights. **Youth Involvement:** Engage young farmers and students in TAK documentation and application, fostering intergenerational knowledge transfer. (9) **Citizen Science Initiatives:** Farmers can act as citizen scientists by: Monitoring environmental changes using TAK indicators. Collecting data on traditional practices (e.g., crop growth patterns, pest outbreaks). (10) **Joint Publications and Intellectual Property: Knowledge Sharing:** Collaborate on publishing research papers or books on TAK, crediting farmers as co-authors. **Intellectual Property Rights (IPR):** Protect farmers' contributions by ensuring equitable benefit-sharing in TAK-based innovations. **Key Benefits of Collaboration: Farmers:** Access to scientific resources, modern tools, and enhanced incomes. **Scientists:** Ground-level insights, practical solutions, and culturally significant data. **Agriculture Sector:** Sustainable and inclusive development, combining innovation with tradition.

18.4 Research Hypotheses on Natural Farming:

There are plenty of theories and hypotheses in ecology which are readily applicable in natural farming. Agricultural scientists should re-evaluate these hypotheses under the conditions of natural farming agroecosystem, to develop new theories for natural farming. Here are some ecological theories which are applicable in natural farming.

18.4.1 General Hypotheses:

Theory of agricultural intensification: (Ester Boserup 1965). Population growth drives agricultural innovation rather than the other way around. As population increases, societies adapt their agricultural practices to enhance productivity, leading to more intensive land use and technological advancements. This counters Thomas Malthus's view that population growth is limited by food supply.

Agricultural location theory: (Johann Heinrich von Thünen in 1826). Agricultural activities are spatially organized based on land cost and transportation expenses. The **Von Thünen Model** proposes concentric rings around a central market, where different types of agriculture cluster based on their profitability and transport costs. The first ring includes high-value crops like dairy, while outer rings feature less perishable products like grains and livestock. This model illustrates the trade-off between land value and transportation costs, influencing farmers' decisions on crop selection and land use.

18.4.2 Hypotheses on Agrobiodiversity Management:

Island Biogeography Theory: (MacArthur and Wilson 1967) Species richness on islands as a balance between immigration and extinction rates. Explains biodiversity patterns in isolated ecosystems like islands, lakes, forest fragments and isolated farms.

Intermediate Disturbance Hypothesis: (Joseph Connell 1978). Biodiversity is highest at intermediate levels of disturbance. **Low Disturbance:** Dominant species outcompete others, reducing diversity. **High Disturbance:** Frequent disturbances eliminate many species. **Intermediate Disturbance:** Allows coexistence of both competitive and resilient species. Helps explain biodiversity in ecosystems like coral reefs, grasslands, forests and natural farming.

Niche Theory: Each species occupies a specific ecological niche, defined by its role in the ecosystem and the resources it uses. **Niche Differentiation:** Species coexist by occupying distinct niches, reducing competition. **Limiting Similarity:** Species with very similar niches cannot coexist indefinitely (competitive exclusion principle). Explains species diversity through resource partitioning and specialization.

Neutral Theory of Biodiversity: (Stephen Hubbell 2001). Species in a community are ecologically equivalent, and biodiversity arises from random processes like speciation, extinction, and dispersal. Explains species richness in tropical forests and coral reefs.

Species–Area Relationship: Larger areas support more species due to greater habitat diversity and resources. Helps design protected areas and conservation strategies.

Keystone Species Hypothesis: Certain species have a disproportionately large impact on ecosystem structure and biodiversity. Removal of a keystone species can lead to ecosystem collapse. Example: honey bees (pollination), earthworms (soil health).

Trophic Cascade Theory: Biodiversity and ecosystem structure are influenced by interactions across trophic levels (e.g., predators, herbivores, plants). **Top-Down Control:** Predators regulate herbivores, maintaining plant diversity. **Bottom-Up Control**: Resource availability influences biodiversity. Explains biodiversity in food webs and ecosystem functioning.

Productivity-Diversity Hypothesis: Biodiversity is influenced by ecosystem productivity (energy or resource availability). **Low Productivity**: Limits species due to resource scarcity. **High Productivity**: Promotes dominance by a few species, reducing diversity. **Intermediate Productivity**: Supports maximum diversity. Explains diversity patterns across ecosystems like deserts and rainforests.

Latitudinal Diversity Gradient: Biodiversity decreases with increasing latitude, peaking in the tropics. Possible explanations include: **Energy Hypothesis**: More sunlight and energy in the tropics support higher diversity. **Evolutionary Time Hypothesis**: Tropics have had stable climates for longer periods, allowing more speciation. **Habitat Heterogeneity Hypothesis**: Tropics offer more complex habitats. Highlights global patterns of biodiversity.

Biodiversity-Stability Hypothesis: Ecosystems with greater biodiversity are more stable and resilient to disturbances. Mechanisms: **Functional redundancy**: Multiple species perform similar roles. **Insurance effect**: Diverse ecosystems can adapt better to changes. Supports the conservation of biodiversity for ecosystem health.

Evolutionary-Ecological Theory: Biodiversity is shaped by the interplay of evolutionary processes (speciation, extinction) and ecological interactions (competition, predation).

Succession Theory: (Frederic Clements, early 20th century). Biodiversity changes over time as ecosystems undergo succession, progressing from pioneer to climax communities. **Primary Succession**: Occurs in areas without previous life (e.g., volcanic regions). **Secondary Succession:** Follows disturbance in existing ecosystems.

Unified Neutral Theory of Biodiversity and Biogeography: (Stephen Hubbell 2001).

Combines neutral theory with island biogeography, suggesting that biodiversity arises from stochastic (random) processes such as speciation, extinction, and dispersal.

Biodiversity-Ecosystem Function Theory (BEF): Biodiversity is directly linked to ecosystem functions like productivity, nutrient cycling, and stability. **Higher Biodiversity:** Increases ecosystem resilience and efficiency. **Functional Diversity:** Emphasizes that it's not just species richness but the variety of roles species play that matter. Used to advocate for biodiversity conservation to maintain ecosystem services critical for human well-being.

Metapopulation Theory: (Richard Levins 1969). Focuses on species populations that are distributed in patches of habitats. Explains the persistence of species in fragmented habitats and informs conservation planning.

Niche Construction Theory: Species actively modify their environment, creating conditions that support other species and enhance biodiversity. Example: Earthworms altering soil properties to support plant growth. Highlights the role of ecosystem engineers in maintaining biodiversity.

Species Coexistence Theory: Explains how multiple species with similar resource requirements coexist in ecosystems without competitive exclusion. Mechanisms: **Temporal variation** (e.g., seasonal resource availability). **Spatial heterogeneity** (e.g., microhabitats). **Non-equilibrium dynamics** (e.g., disturbances). Challenges the idea of stable equilibria and emphasizes dynamic processes in biodiversity maintenance.

Biodiversity Hotspot Theory: (Norman Myers 1988). Biodiversity hotspots are regions with exceptional species richness and endemism, often under threat from human activities.

Energy Hypothesis: Biodiversity is positively correlated with the amount of energy available in an ecosystem. More solar energy or resources (e.g., primary productivity) supports higher species richness. Explains why biodiversity is higher in the tropics than in polar regions.

Janzen-Connell Hypothesis: (Daniel Janzen and Joseph Connell 1970s). High biodiversity in tropical forests is maintained by natural enemies (e.g., herbivores, pathogens) targeting common species. Prevents dominant species from monopolizing resources. Promotes the survival of rare species.

Red Queen Hypothesis: (Leigh Van Valen 1973). Species must continuously evolve to survive because of constant competition, predation, and environmental changes. Highlights the dynamic nature of biodiversity and species interactions.

Energy-Stability-Area (ESA) Theory: (Mark Vellend 2010). Biodiversity is shaped by the interaction of energy availability, environmental stability, and habitat area.

Large, stable areas with abundant energy have higher biodiversity.

Adaptation Radiation Theory: Biodiversity increases when species diversify to exploit new niches, often following environmental changes or colonization of new habitats.

Examples: Darwin's finches in the Galápagos, cichlid fishes in African lakes. Explains bursts of speciation in evolutionary history.

Resource Ratio Theory: (David Tilman 1982). Species coexist by utilizing different ratios of resources (e.g., light, nutrients). Explains species diversity in nutrient-poor and heterogeneous ecosystems.

18.4.3 Hypotheses on Agroecosystem Management

Ecosystem Theory: (Arthur Tansley 1935). An ecosystem is a functional unit consisting of living organisms (biotic) and their physical environment (abiotic), interacting through energy flow and nutrient cycling. Key Processes: Energy transfer through food chains/webs. Matter cycling (carbon, nitrogen, and water cycles). Forms the basis of modern ecological studies and management.

Trophic Dynamics Theory: (Raymond Lindeman 1942). Energy flow in an ecosystem is organized into trophic levels (producers, consumers, decomposers). Only about 10% of energy is transferred to the next trophic level (10% Law). The rest is lost as heat during metabolic processes. Explains food web structures, energy pyramids, and ecosystem productivity.

Top-Down vs. Bottom-Up Control Theory: Ecosystem dynamics are influenced by: **Top-Down Control**: Predators regulate herbivores, which in turn affect plant populations. **Bottom-Up Control**: Resource availability (nutrients, sunlight) determines plant productivity, influencing higher trophic levels. Helps in understanding population control mechanisms and trophic cascades.

Succession Theory: (Frederic Clements 1916) and later refined by Henry Gleason.

Ecosystems change over time through a predictable series of stages, from pioneer species to a stable climax community. **Primary Succession**: Occurs in barren areas without previous life (e.g., after volcanic eruptions). **Secondary Succession**: Occurs in areas with pre-existing life but disturbed (e.g., after fires). Used in habitat restoration and understanding ecosystem resilience.

Ecosystem Services Theory: Ecosystems provide essential services that support human survival and well-being. **Provisioning Services**: Food, water, timber. **Regulating Services**: Climate regulation, pollination, flood control. **Cultural Services:** Recreational, spiritual. **Supporting Services:** Soil formation, nutrient cycling. Underpins conservation efforts and sustainable resource management.

Dynamic Equilibrium Theory: Ecosystems maintain stability through feedback mechanisms despite external disturbances. **Negative Feedback:** Stabilizes the system (e.g., predator-prey dynamics). **Positive Feedback:** Amplifies changes, potentially leading to regime shifts (e.g., deforestation accelerating soil erosion). Explains resilience and tipping points in ecosystems.

Energy Flow Theory: (Eugene Odum 1950s). Ecosystems function through the transfer and transformation of energy, starting from solar input. **Gross Primary Productivity (GPP):** Total energy captured by producers. **Net Primary Productivity (NPP):** Energy remaining after producers use some for respiration. Measures ecosystem productivity and efficiency.

Resilience Theory: (C.S. Holling 1973). Ecosystems can absorb disturbances and still maintain their structure and functions. **Elastic Resilience:** Quick recovery to the original state. **Adaptive Resilience:** Adjusting to new conditions after disturbances. Guides sustainable management and climate change adaptation.

Ecosystem Stability Theory: Ecosystem stability depends on biodiversity and ecological interactions. **Resistance:** Ability to withstand disturbances. **Resilience:** Ability to recover from disturbances. **Functional Redundancy:** Multiple species performing similar roles enhance stability. Highlights the importance of biodiversity for maintaining ecosystem health.

Optimal Foraging Theory: Organisms in an ecosystem behave in ways that maximize energy gains while minimizing energy expenditure. Example: Predators select prey based on abundance and ease of capture.

Thermodynamic Ecosystem Theory: (Howard Odum 1950s). Ecosystems are governed by thermodynamic principles: Energy cannot be created or destroyed. Ecosystems tend toward maximum energy efficiency. Explains energy dissipation and material cycling in ecosystems.

Ecosystem Network Theory: Ecosystems are networks of interacting species and abiotic components. **Strong Links:** Dominant interactions (e.g., predator-prey). **Weak Links:** Contribute to resilience by redistributing energy and resources. Used in understanding food webs and ecosystem complexity.

Patch Dynamics Theory: Ecosystems are mosaics of habitat patches, each at a different stage of succession. **Disturbances:** Create new patches and maintain diversity.

Connectivity: Movement of species between patches supports ecosystem functioning.

Explains biodiversity in fragmented landscapes.

Ecosystem Development Theory: Ecosystems mature over time through:

Increased complexity and biodiversity. Greater energy efficiency. Stabilization of nutrient cycles. Helps in understanding long-term ecological changes and stability.

Anthropogenic Ecosystem Theory: Human activities (e.g., agriculture, urbanization) have transformed natural ecosystems into anthropogenic systems. Ecosystem functions are altered, often reducing biodiversity and resilience. Highlights the need for sustainable development and ecosystem restoration.

Self-Organization Theory: Ecosystems organize themselves into complex, adaptive systems without external control. Feedback loops and species interactions drive this process. Explains the emergence of complex food webs and habitat structures.

Gaia Hypothesis: (James Lovelock 1970s). The Earth functions as a self-regulating system, where living and nonliving components interact to maintain conditions for life. Highlights the interconnectedness of ecosystems on a planetary scale.

Soil Formation Theory: (Hans Jenny 1941). Soil formation depends on five factors: parent material, climate, organisms, topography, and time. Known as the CLORPT model (Climate, Organisms, Relief, Parent material, Time).

Soil Horizons and Pedogenesis: (Vasily Dokuchaev 1879). The concept of soil as a natural body with distinct layers (horizons) formed through processes like leaching, weathering, and organic matter accumulation. Regarded as the "father of soil science."

Soil Fertility Theory: (Justus von Liebig 1840s). The "Law of the Minimum" states that plant growth is limited by the nutrient in shortest supply, emphasizing the importance of soil nutrients like nitrogen, phosphorus, and potassium.

Soil Erosion Theory: (Hugh Hammond Bennett 1930s). Emphasized the dangers of soil erosion and advocated for soil conservation practices to maintain soil productivity.

Soil Microbial Ecology: (Selman Waksman 1940s). Explores the role of microorganisms in soil processes like decomposition, nutrient cycling, and organic matter formation.

Dynamic Equilibrium of Soil (Catena Concept): (Geoffrey Milne 1935). Soils on a slope (catena) vary systematically due to changes in water drainage and topography.

Soil Structure and Aggregation: (H.M. Taylor and E.G. Richards 20th Century). Examines how soil particles bind to form aggregates, which influence aeration, water retention, and root penetration.

Humus Theory: (Albert Howard 1940s). Highlights the role of humus (decomposed organic matter) in maintaining soil fertility and structure.

Soil–Water Retention and Drainage Theory (Field Capacity): Soil Physicists (20th Century). Explains the relationship between soil texture, porosity, and the soil's ability to retain and drain water.

Pedo-transfer Functions: (Jacob Bouma 1989). Predicts soil properties (e.g., hydraulic conductivity) based on easily measurable parameters like texture and organic matter.

Tillage and Compaction Theory: (Edward H. Faulkner 1943, Plowman's Folly). Criticized deep plowing for causing soil compaction and argued for reduced tillage to preserve soil health.

Soil Carbon Sequestration: (Rattan Lal and others 1990s onwards). Storing carbon in soil as organic matter can mitigate climate change while improving soil fertility and structure.

Bioturbation Theory: Bioturbation is the biological reworking of soils by organisms, such as earthworms, through activities like burrowing and feeding. This process significantly impacts agroecosystems by altering nutrient cycling, enhancing microbial activity, and modifying soil structures. It is considered a form of "ecosystem engineering," influencing biodiversity and geochemical gradients. (Filip J R Meysman et al. 2006).

Darwin's Theory on Earthworm Ecology: (Charles Darwin 1881). In his book The Formation of Vegetable Mould through the Action of Worms, Darwin highlighted the ecological role of earthworms in soil formation, nutrient cycling, and plant growth. He emphasized their importance in creating fertile topsoil.

Drilosphere Theory: (Marcel Bouché 1970s). Refers to the soil environment influenced by earthworm activity, including burrows, castings, and the microbial communities associated with their presence. The drilosphere enhances soil aeration, water infiltration, and nutrient availability.

Vermicomposting Theory: (Mary Appelhof 1980s, Ismail Sultan 1997). Earthworms process organic waste into nutrient-rich compost (vermicompost). The theory emphasizes their ability to decompose organic matter, improve soil fertility, and support sustainable waste management.

Ecosystem Engineering by Earthworms: (Clive G. Jones, John H. Lawton, and Moshe Shachak 1994). Earthworms are considered "ecosystem engineers" because they modify soil structure and chemistry, thereby influencing the habitat and availability of resources for other organisms.

Litter-Mixing Hypothesis: (Hendrix et al. 1992). Earthworms incorporate surface litter into the soil, enhancing decomposition rates and nutrient cycling. They also play a role in mixing organic matter with mineral soil layers.

Nutrient Hotspot Theory: (Lavelle et al. 1998). Earthworm casts and burrows act as "hotspots" of microbial activity and nutrient concentration, which significantly enhances soil fertility and plant growth.

Soil Aggregation Theory: (Edwards and Bohlen 1996). Earthworm activity contributes to soil aggregation by binding soil particles with organic matter and microbial secretions, improving soil structure and reducing erosion.

Burrowing and Water Infiltration Theory: (Shipitalo and Le Bayon 2004). Earthworm burrows increase soil porosity, which enhances water infiltration and reduces surface runoff, leading to better soil water retention.

Carbon Sequestration Hypothesis: (Lubbers et al. 2013). Earthworm activity contributes to the stabilization of organic carbon in the soil, supporting carbon sequestration and mitigating climate change.

Trophic Cascade Theory in Soil Ecosystems: (Moore et al. 2004). Earthworms influence trophic cascades in the soil food web by breaking down organic matter and supporting microbial and detritivore populations, which in turn affect plant growth.

Biological Soil Mixing Theory: (Marcel Bouché 1977). Earthworms mix soil layers by their burrowing and feeding activities, which homogenizes soil horizons and enhances nutrient distribution.

Earthworm Ecotypes Theory: Marcel Bouché 1972). Classified earthworms into three ecological groups based on their behavior and habitat: **Epigeic:** Live in surface litter and feed on decaying organic matter. **Endogeic:** Live and feed within the soil. **Anecic:** Create permanent vertical burrows and feed on surface litter.

Soil Food Web Theory: (Diana H. Wall 1999). Earthworms are integral to the soil food web, serving as both decomposers and prey for higher trophic levels, thus maintaining ecosystem stability.

Detoxification of Pollutants by Earthworms: (Edwards and Arancon 2004). Earthworms can degrade organic pollutants and immobilize heavy metals in contaminated soils, contributing to bioremediation efforts.

Earthworm Impact on Microbial Communities: (Lavelle and Spain 2001). Earthworm gut passage enhances microbial diversity and activity, boosting the decomposition of organic matter and nutrient cycling.

Mutualism Theory: (Frank 1885, expanded by Harley and Smith 1983). Mycorrhizae are mutualistic associations between fungi and plant roots, where fungi provide plants with nutrients (e.g., phosphorus) and water, and plants supply fungi with carbohydrates.

Common Mycorrhizal Network (CMN) Theory: (Suzanne Simard and colleagues 1997). Mycorrhizal fungi connect the roots of multiple plants, facilitating the transfer of nutrients, water, and even signaling molecules between plants. This network can enhance plant growth and stress tolerance.

Nutrient Exchange Optimization Theory: (Koide and Elliott 1989). Mycorrhizal fungi and plants optimize nutrient exchange based on supply and demand. Plants allocate more carbohydrates to fungi when nutrient availability in the soil is low.

Ectomycorrhizal Dominance Hypothesis: (Brundrett et al. 1996). Ectomycorrhizal fungi dominate in nutrient-poor soils, particularly in forests, because they are more efficient at mobilizing nutrients from organic matter than arbuscular mycorrhizae.

Functional Diversity Theory: (Read and Perez–Moreno 2003). Mycorrhizal fungi exhibit functional diversity, with different species or strains specializing in mobilizing specific nutrients, enhancing plant diversity and ecosystem resilience.

Stress Gradient Hypothesis (SGH): (Callaway and Walker 1997). The mutualistic role of mycorrhizae is more pronounced under environmental stress (e.g., drought or nutrient deficiency), where they help plants cope by improving water and nutrient uptake.

Carbon Cost-Benefit Theory: (Johnson et al. 1997). Mycorrhizal associations are sustained when the carbon cost to the plant is outweighed by the benefits in terms of nutrient acquisition and growth.

Nutrient Mining Theory: (Smith and Read 2008). Mycorrhizal fungi extend their hyphae into the soil to access nutrients beyond the root depletion zone, effectively "mining" phosphorus, nitrogen, and other essential nutrients.

Mycorrhizal Symbiosis and Plant Defense Theory: (Gianinazzi et al. 1995). Mycorrhizae enhance plant resistance to pathogens by improving nutrient status, stimulating systemic resistance, and competing with soil-borne pathogens.

Mycorrhizal Resource Partitioning Theory: (Bever et al. 2001). Different mycorrhizal species colonize plants based on nutrient availability, soil type, and plant species, leading to partitioning of resources within the ecosystem.

Mycorrhizal Dependency Theory: (Gerdemann 1968). Some plants are highly dependent on mycorrhizae for survival and growth, especially in nutrient-poor soils, while others show less dependency.

Hydraulic Redistribution Theory: (Egerton-Warburton et al. 2007). Mycorrhizal fungi help redistribute water from deeper soil layers to shallower ones, benefiting both the host plant and neighboring plants during drought.

Soil Carbon Sequestration Hypothesis: (Rillig et al. 2001). Mycorrhizal fungi contribute to soil carbon storage by producing glomalin, a glycoprotein that stabilizes soil aggregates and enhances carbon retention.

Host-Specificity and Co-Evolution Theory: (Bruns and Taylor 1996). Mycorrhizal fungi and their host plants exhibit varying degrees of specificity, with some associations resulting from co-evolution over millions of years.

Dual Mycorrhizal Associations Theory: (Tedersoo et al. 2010). Some plants can associate with both ectomycorrhizal and arbuscular mycorrhizal fungi, gaining a competitive advantage by accessing multiple nutrient sources.

Global Distribution and Biogeography Theory: (Tedersoo and Smith 2013). Mycorrhizal fungi distribution patterns are influenced by climatic, soil, and plant community factors, shaping global ecosystem functioning.

Functional Equilibrium Model: (Johnson et al. 2003). Plants allocate more resources to roots and mycorrhizae under nutrient limitations, ensuring optimal nutrient acquisition relative to carbon investment.

Arbuscular Mycorrhizal Pathway Theory: (Smith and Smith 2011). The arbuscular mycorrhizal pathway (AMP) facilitates direct nutrient uptake into the plant root through fungal structures like arbuscules.

Rhizosphere Competence Theory: (Hiltner 1904). PGPR thrive in the rhizosphere due to their ability to colonize and compete for nutrients and space effectively, forming beneficial associations with plant roots.

Plant–Microbe Mutualism Theory: (Paul and Clark 1989). PGPR establishes mutualistic relationships with plants by providing growth–promoting substances or nutrients in exchange for root exudates as a carbon source.

Biological Nitrogen Fixation Theory: (Beijerinck 1901 and Burris 1974). Nitrogen-fixing PGPR (e.g., Azospirillum, Rhizobium) converts atmospheric nitrogen into ammonia, enhancing plant nitrogen availability and growth.

Phytohormone Production Theory: (Dobbelaere et al. 2003). PGPR produces phytohormones like auxins (e.g., indole-3-acetic acid), cytokinins, and gibberellins, which stimulate root elongation, cell division, and overall plant growth.

Siderophore Production Theory: (Neilands 1981). PGPR secrete siderophores that chelate iron in the soil, making it available to plants while limiting its availability to pathogenic microbes.

Induced Systemic Resistance (ISR) Theory: (Van Loon et al. 1998). PGPR trigger ISR in plants, enhancing their defense mechanisms against pathogens and abiotic stresses without directly targeting the pathogen.

Antibiotic Production Theory: (Thomashow and Weller 1988). PGPR produce antibiotics (e.g., phenazines, pyoluteorin) to suppress pathogenic microorganisms in the rhizosphere, promoting plant health.

ACC Deaminase Theory: (Glick et al. 1998). PGPR producing ACC deaminase lower ethylene levels in plants by breaking down its precursor (1-aminocyclopropane-1-carboxylic acid), reducing stress-induced growth inhibition.

Phosphate Solubilization Theory: (Kucey et al. 1989). PGPR solubilizes inorganic phosphate from the soil, converting it into forms accessible to plants, enhancing phosphorus uptake and growth.

Quorum Sensing and Communication Theory: (Fuqua et al. 1994). PGPR use quorum sensing to coordinate activities like biofilm formation, antibiotic production, and colonization, ensuring effective plant-microbe interactions.

Stress Alleviation Theory: (Yang et al. 2009). PGPR mitigate abiotic stresses (e.g., drought, salinity) by producing osmoprotectants, antioxidants, and enzymes that enhance plant stress tolerance.

Soil Remediation and Biodegradation Theory: (Anderson et al. 1993). PGPR degrades pollutants (e.g., hydrocarbons, heavy metals) in contaminated soils, improving soil quality and enabling plant growth in degraded environments.

Biofilm Formation Theory: (Costerton et al. 1995). PGPR form biofilms on plant roots, enhancing colonization, nutrient exchange, and protection against desiccation and pathogens.

Nutrient Cycling Theory: (Vessey 2003). PGPR contributes to nutrient cycling by mineralizing organic matter and mobilizing nutrients (e.g., potassium, zinc), making them available to plants.

PGPR-Strain Specificity Theory: (Berg et al. 2009). Different PGPR strains exhibit specificity for particular plant species or genotypes, optimizing their growth-promoting effects under specific environmental conditions.

Reduction of Heavy Metal Toxicity Theory: (Rajkumar et al. 2012). PGPR alleviates heavy metal toxicity in plants by sequestration, reduction, or immobilization of heavy metals in the rhizosphere.

Synergistic PGPR Communities Theory: (Compant et al. 2010). PGPR often function as part of microbial consortia, where multiple species work synergistically to enhance plant growth and resilience.

Co-Evolution of PGPR and Plants Theory: (Oldroyd and Leyser 2020). PGPR and plants have co-evolved, with plants selecting beneficial rhizobacteria through root exudates and bacteria adapting to colonize specific plant hosts.

Hydrogen Cyanide (HCN) Production Theory: (Voisard et al. 1989). Some PGPR produce HCN, a secondary metabolite that suppresses soil-borne pathogens, contributing to plant protection.

Rhizosphere Priming Effect Theory: (Kuzyakov et al. 2000). PGPR enhances the decomposition of soil organic matter through enzyme production and microbial interactions, making nutrients available to plants.

Bioprecipitation Theory: Bioprecipitation refers to the process by which microorganisms, particularly ice-nucleating bacteria, influence precipitation patterns. Proposed by David Sands in the 1970s, it involves land plants releasing aerosols that contain these bacteria, which facilitate ice formation in clouds at warmer temperatures than pure water would allow. This mechanism is crucial for rain and snow, as it enhances cloud formation and precipitation efficiency. Bacteria like *Pseudomonas syringae* are notable for their role in this process, acting as catalysts for ice nucleation (Morris CE et al. 2014)

Watershed Management Theory: (USDA Soil Conservation Service 1935). Focuses on managing the entire watershed to conserve rainwater by reducing runoff, increasing groundwater recharge, and controlling soil erosion.

Integrated Water Resource Management (IWRM) Theory: (GWP 2000). Promotes the coordinated management of rainwater, surface water, and groundwater to optimize water use and ensure sustainable resource development.

Water Budgeting Theory: (Thornthwaite and Mather 1955). Advocates balancing water inputs (e.g., rainfall, harvesting) with outputs (e.g., evaporation, transpiration) to optimize water use and conservation.

Zero Runoff Theory: (C. J. Perry 2001). Encourages techniques to ensure all rainwater is captured and either stored or infiltrated into the soil, eliminating surface runoff.

Hydrological Cycle and Water Balance Theory: (Horton 1933). Understanding the hydrological cycle helps in designing rainwater harvesting systems to balance precipitation, evaporation, infiltration, and runoff.

Rainwater Harvesting Potential Theory: (Gould and Nissen-Petersen 1999). Defines the potential for rainwater harvesting based on rainfall intensity, collection surface area, and system efficiency.

Catchment Area Management Theory: (Falkenmark 1986). Suggests managing catchment areas (e.g., rooftops, land surfaces) to maximize rainwater collection efficiency and reduce water loss.

Groundwater Recharge Theory: (Todd and Mays 2005). Rainwater harvesting systems can help recharge aquifers by allowing water to percolate into the ground, ensuring long-term water availability.

Check Dam and Contour Bunding Theory: (Fertilizer Association of India (FAI), 1970s). Construction of check dams and contour bunds slows down water flow, enabling storage and infiltration, which reduces erosion and increases groundwater recharge.

Ecohydrology Theory: (Zalewski 2000). Integrates ecological and hydrological processes to optimize water storage, reduce runoff, and enhance water availability for ecosystems.

Roof Water Harvesting (RWH) Theory: (Gould and Petersen 1999). Capturing and storing rainwater from rooftops for domestic or agricultural use is a sustainable way to supplement water needs.

Rainwater Quality Management Theory: (Fewkes and Butler 2000). Ensures harvested rainwater meets quality standards by reducing contamination through proper design and maintenance of collection systems.

Percolation Tank Theory: (Indian Council of Agricultural Research ICAR). Percolation tanks store rainwater and allow it to gradually infiltrate into the soil, improving groundwater levels.

Sponge City Concept: (Yu Kongjian 2000s, China). Urban areas are designed to act like sponges, capturing, storing, and purifying rainwater to reduce urban flooding and enhance water availability.

Rainwater Harvesting System Efficiency Theory: (Fewkes 1999). Focuses on maximizing the efficiency of rainwater harvesting systems by optimizing storage design, collection surfaces, and filtration mechanisms.

In-Situ Water Conservation Theory: (Indian Council of Agricultural Research ICAR). Emphasizes conserving rainwater where it falls by using techniques like mulching, contour plowing, and deep tillage.

Runoff Coefficient Theory: (Chow et al. 1988). Defines the fraction of rainfall that becomes runoff, helping in the design of rainwater harvesting systems.

Socio-Hydrology Theory: (Sivapalan et al. 2012). Studies the interactions between human activities and the hydrological cycle to design sustainable water management practices, including rainwater harvesting.

Circular Economy of Water Theory: (Ellen MacArthur Foundation 2015). Emphasizes the reuse and recycling of rainwater to close the water loop and reduce dependency on conventional water sources.

Rainwater Harvesting for Resilience Theory: (Rockström et al. 2009). RWH increases resilience to climate variability by providing supplementary water for agricultural, domestic, and industrial use during dry spells.

18.4.4 Hypotheses on Honey Bee and Pollination:

Theory of Kin Selection: (William D. Hamilton 1964). Explains altruistic behavior in honey bee colonies. Worker bees, although sterile, help the queen (their mother) reproduce because it ensures the propagation of shared genes.

Waggle Dance Theory: (Karl von Frisch 1944). Honey bees communicate the location of food sources through a "waggle dance," which conveys the direction and distance relative to the sun. Karl von Frisch won a Nobel Prize for this discovery.

Superorganism Theory: (William Morton Wheeler 1911, expanded by E.O. Wilson and Bert Hölldobler 1990). A honey bee colony functions as a single "superorganism," where individual bees act like cells of a body, collectively ensuring the survival and reproduction of the colony.

Optimal Foraging Theory: (Eric L. Charnov 1976) and applied to bees by numerous researchers. Honey bees optimize their foraging strategies by balancing energy expenditure with the nutritional gain from nectar and pollen collection.

Nest-Site Selection Theory: (Thomas D. Seeley and colleagues 1980s). When swarming, honey bees use a democratic process to select the best nest site. Scout bees present options through waggle dances, and the colony collectively decides based on consensus.

Division of Labor Theory: (Charles Darwin 1859, as a part of natural selection) and later expanded by R.E. Page and G.E. Robinson (1991). Worker bees in a colony perform specific tasks based on age (age polyethism) and physiological condition, ranging from nursing to foraging.

Pollination Ecology Theory: (Charles Robertson 1890s), expanded by David W. Inouye and others. Highlights the critical role of honey bees in pollination networks, emphasizing coevolution between bees and flowering plants.

Colony Collapse Disorder (CCD) Hypotheses: Various researchers (2006 onwards). Investigates the causes of sudden hive abandonment by worker bees. Factors include pesticides (e.g., neonicotinoids), pathogens, habitat loss, and climate change.

Self-Organization in Honey Bees: (Deborah Gordon 1999, applied to honey bees by Seeley). Complex behaviors in honey bee colonies emerge from simple individual interactions, such as pheromone signaling and task allocation.

Queen Pheromone Theory: (Mary Jane West-Eberhard 1987). The queen releases specific pheromones that regulate colony cohesion, worker reproduction, and task behaviors.

Sexual Selection in Drone Congregation Areas: (Robin Moritz and others 1980s). Drones congregate at specific locations to mate with queens. Sexual selection ensures genetic diversity and colony fitness.

Haplodiploidy and Eusociality Theory: (William D. Hamilton 1964). The haplodiploid genetic system (males are haploid, females are diploid) in honey bees promotes eusocial behavior by enhancing genetic relatedness among sisters.

Honey Bee Thermoregulation Theory: (Bernd Heinrich 1970s). Honey bees regulate hive temperature by clustering, fanning, or evaporating water to maintain optimal conditions for brood development.

Behavioral Plasticity Theory: (G.E. Robinson and others 1990s). Worker bees exhibit plasticity in their behavior, allowing them to switch tasks based on colony needs, environmental factors, or individual aging.

Honey Bee Memory and Learning: (Martin Giurfa 1990s). Honey bees demonstrate remarkable learning abilities, such as associating floral patterns, scents, and colors with rewards (nectar).

Resilience Theory in Bee Ecology: (Buzz Holling 1973, applied to bee populations in recent years). Examines the resilience of honey bee populations to environmental stressors, such as habitat loss and pesticide exposure.

Pollination Syndrome Theory: (Darwin 1862, expanded by Faegri and van der Pijl 1979). Plants evolve specific floral traits (e.g., color, shape, scent) to attract specific pollinators such as bees, birds, bats, or wind, optimizing pollination efficiency.

Co-Evolution Theory: (Darwin 1859). Pollinators and flowering plants co-evolve traits that enhance mutualistic interactions. For example, long nectar spurs in flowers and long proboscises in pollinators.

Optimal Foraging Theory: (Pyke 1978). Pollinators choose flowers that maximize their energy gain relative to the energy expended, influencing plant-pollinator interactions and pollination efficiency.

Floral Constancy Theory: (Waser 1986). Pollinators tend to visit flowers of the same species within a foraging bout, ensuring more effective pollen transfer.

Outbreeding and Genetic Diversity Theory: (Stebbins 1970). Cross-pollination facilitated by pollinators promotes genetic diversity and adaptability in plant populations.

Self-Incompatibility Theory: (East and Mangelsdorf 1925). Many plants have mechanisms to prevent self-pollination (e.g., genetic self-incompatibility), encouraging cross-pollination by biotic agents.

Pollination Ecology Theory: (Heinrich 1979, Real 1983). Examines the interactions between plants and their pollinators in an ecological context, including factors like competition, resource availability, and community dynamics.

Specialist vs. Generalist Pollination Theory: (Johnson and Steiner 2000). Plants may rely on specific pollinators (specialists) or a broad range of pollinators (generalists) depending on their evolutionary strategy and ecological conditions.

Pollinator Limitation Theory: (Larson and Barrett 2000). Reproductive success in some plants is limited by the availability or efficiency of pollinators, influencing floral evolution and reproductive strategies.

Pollen Transfer Efficiency Theory: Harder and Thomson 1989). Efficiency of pollen transfer is influenced by floral morphology, pollinator behavior, and environmental factors, affecting plant reproductive success.

Wind and Water Pollination (Abiotic Pollination) Theory: (Kerner von Marilaun 1895). Abiotic pollination mechanisms (e.g., wind, water) evolve in plants where biotic pollinators are scarce or inefficient.

Mimicry Pollination Theory: (Schiestl and Ayasse 2001). Some plants mimic the appearance or scent of other species (e.g., flowers mimicking female insects) to attract pollinators without providing rewards.

Pollinator Decline Theory: (Kearns, Inouye, and Waser 1998). Declines in pollinator populations due to habitat loss, pesticides, and climate change threaten global plant reproduction and food security.

Nectar Robbing and Its Effects Theory: (Irwin and Brody 1998). Some organisms steal nectar without pollinating the plant, which can reduce plant reproductive success and affect pollination dynamics.

Geitonogamy and Pollinator Behavior Theory: (Lloyd and Schoen 1992). Pollinators often visit multiple flowers on the same plant, increasing self-pollination (geitonogamy) and potentially reducing genetic diversity.

Buzz Pollination (Sonication) Theory: (Buchmann 1983). Some plants require pollinators (e.g., bees) to vibrate their flight muscles to release pollen, known as buzz pollination.

Pollination Efficiency and Floral Resource Dynamics: (Zimmerman and Pyke 1988). Pollination success is influenced by the quantity and quality of floral rewards (nectar, pollen) and how pollinators interact with these resources.

Trapline Foraging Theory: (Thomson et al. 1986). Pollinators follow specific foraging routes (traplines) to maximize energy efficiency, which can influence pollination patterns in plant populations.

Pollen–Ovule Ratio Theory: (Cruden 1977). Plants with higher pollen-ovule ratios often rely on biotic pollinators, while those with lower ratios may utilize abiotic pollination.

Climate Change and Pollination Theory: (Hegland et al. 2009). Climate change alters the timing and behavior of pollinators, potentially disrupting plant-pollinator interactions and reproductive success.

18.4.5 Hypotheses on Natural Control of Insect Pests, Diseases and Weeds

Theory of Natural Control: (F.P. Wiltshire 1947). Populations of pests and their natural enemies are regulated naturally by environmental factors, without human intervention. It forms the basis of biological control concepts.

Augmentation Theory: (Paul DeBach 1964). Involves augmenting the population of natural enemies through mass release or by providing conditions favorable for their reproduction. Two types: **Inoculative Release:** Releasing a small number of biological control agents that reproduce and establish control over time. **Inundative Release:** Releasing a large number of agents to immediately reduce pest populations.

Ecological Balance Theory: (Charles Elton 1958). Biological control is based on maintaining ecological balance between pests and their natural enemies, ensuring neither overpopulation nor extinction.

Enemy Release Hypothesis (ERH): (Charles S. Elton 1958), expanded by several researchers in later years. Non-native species often become invasive because they escape their natural enemies in the new environment, leading to uncontrolled population growth.

Classical Biological Control Theory: (Harry Smith and others 1930s-1940s). Involves the introduction of a pest's natural enemy from its native habitat into a new area where the pest has become a problem.

Pest Resurgence Theory: (Stern et al. 1959). The use of chemical pesticides can disrupt natural enemy populations, leading to a resurgence of pest populations that were previously under biological control.

Apparent Competition Theory: (Holt and Lawton 1993). Occurs when two prey species share a common predator; an increase in one prey population indirectly harms the other by boosting predator populations.

Host–Parasitoid Interaction Models: (Robert May 1973). Explains population dynamics between pests (hosts) and their natural enemies (parasitoids), emphasizing stability through periodic oscillations.

Trophic Cascade Theory: (Hairston, Smith, and Slobodkin 1960). Top predators (natural enemies) indirectly benefit plants by reducing herbivore populations, thus maintaining ecosystem health.

Functional Response Theory: (C.S. Holling 1959). Describes the rate at which a predator consumes prey as prey density increases. It has three types: **Type I:** Linear increase in predation. **Type II:** Predation rate slows as prey density increases (handling time limitation). **Type III:** S-shaped curve due to learning or prey switching by predators.

Density-Dependent Regulation Theory: (David Lack 1954). Natural enemies regulate pest populations more effectively as pest density increases, helping maintain balance in ecosystems.

Integrated Pest Management (IPM) Theory: (Stern et al. 1959). Combines biological control with cultural, mechanical, and chemical methods to manage pests while minimizing environmental impact.

Predator-Prey Oscillation Theory: (Alfred Lotka and Vito Volterra 1926). Describes predator-prey population cycles and their dynamic equilibrium, providing insights into biological control dynamics.

Self-Regulation Theory: (Andrewartha and Birch 1954). Populations self-regulate through density-dependent factors like competition and predation, minimizing the need for external interventions.

Invasion Ecology Theory: (Richard H. Groves and J.J. Burdon 1986). Addresses the biological control of invasive species by reintroducing their native natural enemies.

Competition Theory: Gause's Competitive Exclusion Principle: Two species competing for the same limited resources cannot coexist indefinitely; one species will outcompete and exclude the other unless they adapt through niche differentiation. **Resource Partitioning:** Coexisting species evolve to exploit different resources or niches to reduce competition.

Mutualism Theory: Coevolution: Mutualistic interactions drive the co-evolution of species, as they adapt to enhance mutual benefits (e.g., pollinators and flowering plants). **Facilitation Model:** Suggests that mutualistic species improve the conditions for each other, promoting survival and growth in challenging environments.

Predation and Herbivory Theories: Optimal Foraging Theory: Predators or herbivores maximize their energy gain per unit of foraging effort, influencing prey or plant species' defensive strategies. **Lotka-Volterra Predator-Prey Model:** Describes oscillating population dynamics where predator and prey populations regulate each other over time.

Parasitism Theory: Red Queen Hypothesis: Suggests a continuous evolutionary arms race between hosts and parasites, with hosts evolving defenses and parasites counter-evolving. **Virulence-Transmission Trade-Off Hypothesis:** Parasites evolve to balance their virulence with their ability to transmit to new hosts, ensuring their survival.

Facilitation and Amensalism: Facilitation Theory: Positive interactions, where one species improves conditions for another (e.g., nurse plants improving soil for seedlings). **Allelopathy:** Chemical inhibition by one species that negatively affects another, common in plant competition.

Ecological Guild and Keystone Species Theory: Ecological Guild: Groups species that exploit similar resources, potentially leading to competition or mutualistic dependencies. **Keystone Species Hypothesis:** Some species have disproportionately large effects on community structure and interspecific interactions.

Symbiotic Relationships Framework: Symbiosis: Long-term interactions such as mutualism, commensalism, and parasitism that evolve specific adaptive strategies between species. **Enemy Release Hypothesis:** Invasive species often thrive in new environments because their natural predators or competitors are absent, altering native interspecific interactions.

Trophic Cascade Theory: Describes how changes at one trophic level (e.g., top predators) ripple through the food web, influencing interspecific relationships across multiple levels (e.g., predators indirectly benefiting producers by controlling herbivores).

Game Theory in Ecology: Applies to interspecific interactions like cooperation, cheating in mutualisms, or predator-prey dynamics: **Evolutionary Stable Strategies (ESS):** Strategies that, when adopted by a population, cannot be invaded by alternative strategies. **Prisoner's Dilemma in Mutualism:** Explores conditions under which species cooperate or cheat in mutualistic relationships.

Stress Gradient Hypothesis: Predicts that facilitative interactions are more common in harsh environments (e.g., deserts, high altitudes), whereas competitive interactions dominate in less stressful environments.

Batesian and Müllerian Mimicry Theories: Batesian Mimicry: Harmless species evolve to imitate harmful ones, benefiting from predator avoidance. **Müllerian Mimicry:** Multiple harmful species converge on similar warning signals to enhance predator learning.

18.4.6 Hypotheses on Marketing

The 4Ps of Marketing (Marketing Mix): (E. Jerome McCarthy 1960). Successful marketing is based on four key elements: **Product:** What is being offered. **Price:** The cost to the customer. **Place:** Distribution channels. **Promotion:** Advertising and sales efforts.

Maslow's Hierarchy of Needs: (Abraham Maslow 1943). Consumers' purchasing behavior is driven by their needs, categorized into five levels: **Physiological:** Basic survival (e.g., food, water). **Safety:** Security and protection. **Social:** Belonging and relationships. **Esteem:** Recognition and status. **Self-Actualization:** Personal growth and fulfillment.

AIDA Model: (E. St. Elmo Lewis 1898). Describes the stages of consumer behavior during the buying process: **Awareness:** Capturing attention. **Interest:** Generating curiosity. **Desire:** Creating a want for the product. **Action:** Prompting purchase or engagement.

Porter's Five Forces: (Michael Porter 1979). Analyzes competitive forces in an industry to guide marketing strategies: **Competitive Rivalry:** Intensity of competition. **Threat of New Entrants:** Ease of entering the market. **Bargaining Power of Suppliers:** Their influence on pricing. **Bargaining Power of Buyers:** Customers' influence on the market. **Threat of Substitutes:** Availability of alternative solutions.

SWOT Analysis: Evaluates internal and external factors affecting a business: **Strengths:** Internal advantages. **Weaknesses:** Internal disadvantages. **Opportunities:** External favorable conditions. **Threats:** External risks or challenges.

Consumer Behavior Theory: Explains how consumers make purchasing decisions based on psychological, social, and cultural influences. **Psychological Factors:** Perception, motivation, learning. **Social Factors:** Family, peer groups. **Cultural Factors:** Beliefs, values, customs.

Relationship Marketing Theory: Focuses on building long-term relationships with customers rather than one-time transactions. **Key Strategies:** Loyalty programs, personalized communication, after-sales support.

Segmentation, Targeting, and Positioning (STP) Model: A framework for identifying and serving the right customers: **Segmentation:** Dividing the market into distinct groups. **Targeting:** Selecting the most profitable segments. **Positioning:** Crafting a unique value proposition for the target market.

Push and Pull Strategy: Two approaches to demand creation: **Push Strategy:** Promoting products through distribution channels (e.g., retailer incentives). **Pull Strategy:** Generating direct demand from consumers (e.g., advertising).

Blue Ocean Strategy: (W. Chan Kim and Renée Mauborgne 2005). Focuses on creating new, uncontested market spaces ("Blue Oceans") instead of competing in existing saturated markets ("Red Oceans").

Diffusion of Innovation Theory: (Everett Rogers 1962). Explains how new products or technologies are adopted over time: **Innovators → Early Adopters → Early Majority → Late Majority → Laggards.**

The Long Tail Theory: (Chris Anderson 2004). Focuses on niche markets rather than mainstream, high-demand products. Online platforms benefit by offering a wide variety of niche products. Guides e-commerce and digital marketing strategies.

Hierarchy of Effects Model: (Robert Lavidge and Gary Steiner 1961). Describes the stages consumers go through before purchasing: **Awareness → Knowledge → Liking → Preference → Conviction → Purchase.**

Game Theory in Marketing: Businesses make strategic decisions by anticipating competitors' actions. Examples: Pricing wars, promotional strategies.

Viral Marketing Theory: Marketing messages spread exponentially through social sharing, like a virus. Relies on emotional content and user-generated sharing.

Resource-Based View (RBV): A company's competitive advantage lies in its unique resources and capabilities (e.g., brand reputation, patents, skilled workforce). Focuses on leveraging internal strengths to dominate the market.

Customer Lifetime Value (CLV): Focuses on maximizing the value a customer brings to a business over the entire relationship. Encourages long-term retention and high-value customer engagement.

Lovemarks Theory: (Kevin Roberts 2004). Brands should evoke loyalty beyond reason by creating emotional connections. Combines mystery, sensuality, and intimacy to foster strong brand love.

18.5 Characteristics of a Model Natural Farm for Research:

It is absolutely essential to have a model natural farm, to initiate any kind of research experiment on natural farming. A conventional agricultural farm is not at all suitable for conducting research on natural farming, otherwise the data will not be valid. Here are the main characteristics of an ideal natural farming agroecosystem:

18.5.1 Optimum Agrobiodiversity

Agrobiodiversity in a natural farming agroecosystem refers to the variety and variability of plants, animals, microorganisms, and ecosystems used directly or indirectly for agricultural production. Its characteristics emphasize sustainability, resilience, and ecological harmony. Below are its key features: (1) **High Species Diversity: Crop Diversity**: Cultivation of multiple crop species and varieties, including staple crops, vegetables, fruits, legumes, and medicinal plants. Polyculture systems, intercropping, and crop rotations are common. **Livestock Diversity**: Inclusion of diverse livestock breeds adapted to local conditions, contributing to nutrient cycling and farm productivity. **Microbial Diversity**: A wide range of beneficial soil microorganisms, including bacteria, fungi, and mycorrhizae, that support nutrient cycling and plant health. (2) **Genetic Diversity: Indigenous Varieties**: Use of local crop varieties and livestock breeds that are well-suited to the region's climatic and soil conditions. Maintenance of landraces and heirloom seeds to preserve genetic traits. **Climate Resilience**: Genetically diverse crops and animals are better adapted to withstand drought, pests, diseases, and climate variability. (3) **Functional Diversity: Ecological Roles**: Integration of species with complementary functions (e.g., nitrogen-fixing legumes, pest-repelling plants, and shade-providing trees). Use of companion planting to improve growth, deter pests, and enhance soil fertility. (4) **Habitat Diversity: Multiple Niches**: Diverse habitats within the farm, such as crop fields, orchards, water bodies, hedgerows, and forest patches. Microhabitats like mulch layers, compost pits, and wild plant areas support biodiversity. **Wildlife Conservation**: Preservation of wild plant and animal species that coexist with farming practices, enhancing ecosystem services like pollination and pest control.

(5) **Integration of Crop-Livestock Systems: Nutrient Cycling:** Livestock provides manure for soil enrichment, while crop residues serve as fodder for animals. **Complementary Roles:** Integration of poultry, goats, cows, or bees enhances productivity and biodiversity. (6) **Temporal Diversity: Seasonal Crops:** Cultivation of crops suited to different seasons to ensure year-round productivity and resource use. **Crop Rotations:** Rotational planting reduces pest and disease buildup and enhances soil fertility. (7) **Ecosystem Services: Pollination:** Diverse plant species attract and sustain pollinators, ensuring efficient crop production. **Pest and Disease Control:** Natural predators and parasitoids thrive in biodiverse systems, regulating pest populations. **Soil Fertility:** Leguminous plants and organic matter recycling maintain and enhance soil fertility. (8) **Local Adaptation: Cultural Relevance:** Practices align with local traditions, making them easier for farmers to adopt and sustain. **Region-Specific Species:** Selection of crops and livestock adapted to local environmental conditions. (9) **Zero Dependence on External Inputs: Self-Sufficiency:** Reliance on farm-produced seeds, fertilizers (compost, manure), and pest control methods reduces external input dependence. **Resource Recycling:** Organic waste is efficiently recycled within the system to sustain biodiversity. (10) **Stability and Resilience: Ecological Balance:** Diverse agroecosystems are less vulnerable to pests, diseases, and climate extremes. **Risk Mitigation:** Diversity in crops and livestock spreads risk, ensuring some level of productivity even in adverse conditions. (11) **Ethical and Sustainable Practices: Animal Welfare:** Ethical treatment of livestock, avoiding overexploitation, and maintaining natural behaviors. **Zero Chemical Use:** Biodiversity thrives due to the absence of synthetic pesticides, herbicides, and fertilizers. (12) **Cultural and Nutritional Benefits: Cultural Heritage:** Conservation of traditional knowledge and farming practices associated with biodiversity. **Nutritional Security:** Diverse crops

and livestock provide a variety of nutrients, promoting a balanced diet for farming families and local consumers.

18.5.2 Climax Agroecosystem:

The climax ecosystem of a natural farming agroecosystem represents a state of ecological balance and maximum sustainability achieved through natural processes and minimal human intervention. Below are its key characteristics: (1) **Biodiversity and Stability: High Biodiversity**: Rich diversity of plant, animal, insect, and microbial life, promoting ecological resilience. Includes a mix of perennial and annual crops, wild plants, and natural predators. **Ecological Stability**: System is self-regulating and less prone to disruptions from pests, diseases, or environmental stress. Stable population dynamics among species due to balanced predator-prey relationships. (2) **Soil Health: Rich Organic Matter**: High levels of humus and organic carbon from continuous decomposition of organic material. Soil structure is porous, allowing efficient water infiltration and root penetration. **Microbial Activity**: Thriving soil microbiome, including bacteria, fungi, and earthworms, facilitates nutrient cycling and plant health. **Nutrient Cycling**: Efficient recycling of nutrients through composting, crop residues, and animal manure. Zero dependence on external inputs like synthetic fertilizers. (3) **Energy Efficiency: Closed Energy System**: Energy flow is efficiently captured through photosynthesis and recycled within the system. Minimal energy loss as waste is naturally decomposed and reintegrated. **Low Input Requirements**: System relies on renewable inputs (e.g., sunlight, rainfall, biomass) rather than fossil fuels. (4) **Water Management: Efficient Water Use**: High water retention capacity due to mulching, organic matter, and natural irrigation methods. Reduced runoff and soil erosion through diverse plant cover and contour farming. **Natural Hydrology**: Restoration of natural water cycles, including groundwater recharge and reduced evaporation. (5) **Pest and Disease Regulation: Natural**

Pest Control: Presence of predatory insects, birds, and beneficial microorganisms controls pest populations. Polyculture and crop rotation reduce the risk of pest and disease outbreaks. **Resilient Plant Health:** Plants are naturally robust due to nutrient-rich soil and diverse agroecosystem support. (6) **Carbon Sequestration: Soil Carbon Storage:** Increased organic matter sequesters significant amounts of carbon in the soil. **Vegetative Carbon Storage:** Diverse vegetation acts as a carbon sink, contributing to climate mitigation. (7) **Minimal Human Intervention: Self-Sustaining Systems:** Ecosystem processes like pollination, nutrient cycling, and pest regulation occur without external aid. **Low Maintenance:** Farmers act as facilitators rather than managers, guiding the system with minimal disturbance. (8) **Agroecological Balance: Crop-Animal Integration:** Livestock contributes to nutrient cycling through manure while benefiting from crop residues. **Symbiotic Relationships:** Companion planting and natural interdependencies optimize resource use and reduce competition. (9) **Productivity and Profitability: Sustained Yields:** Stable and consistent yields due to enhanced soil health and biodiversity. **High-Quality Produce:** Nutrient-dense, chemical-free crops with better taste and market value. (10) **Ecosystem Services: Pollination:** Abundant pollinators ensure effective crop reproduction. **Climate Resilience:** System withstands droughts, floods, and other climatic shocks due to its diversity and regenerative capacity. **Cultural and Aesthetic Value:** The ecosystem often aligns with traditional farming practices, enhancing cultural heritage.

18.5.3 Model Marketing System:

A model marketing system for a small-scale natural farm aims to create an efficient and sustainable way to sell products while maintaining transparency and emphasizing the farm's values. Here's a structured approach: (1) **Define Target Market: Local Consumers:** Prioritize customers interested in fresh, natural, and locally grown produce.

Urban Buyers: Target urban families seeking organic and healthy food options. **Institutions:** Supply schools, restaurants, and organic stores with natural farm products. (2) **Diversify Product Range: Primary Products:** Fresh fruits, vegetables, grains, herbs, and dairy (if applicable). **Value-Added Products:** Jams, pickles, sauces, and dried herbs. Cold-pressed oils, natural soaps, or organic fertilizers. **Seasonal Offerings:** Focus on seasonal produce to ensure quality and variety. (3) **Establish a Brand Identity: Farm Name and Logo:** Create a recognizable name and logo reflecting the farm's values (e.g., sustainability, purity). **Storytelling:** Share the farm's journey and commitment to natural farming through brochures, websites, and social media. **Certifications:** Obtain organic/natural farming certifications to enhance credibility. (4) **Build Sales Channels:** (a) **Direct Sales: Farm Gate Sales:** Allow customers to purchase directly from the farm. **Farmers' Markets:** Set up stalls in local markets or weekly organic bazaars. **Community-Supported Agriculture (CSA):** Offer subscription-based models where customers receive weekly produce boxes. (b) **Digital Platforms: Website:** Develop an e-commerce site for online ordering and home delivery. **Social Media:** Use platforms like Instagram and Facebook for promotions, storytelling, and direct sales. **Local Delivery Apps:** Partner with delivery services or create your own delivery network. (c) **Partnerships:** Collaborate with organic stores, eco-conscious restaurants, or wellness centers for regular supplies. (5) **Leverage Marketing Strategies:** (a) **Customer Engagement: Farm Visits:** Invite customers for guided tours to see how produce is grown naturally. **Workshops:** Conduct sessions on natural farming, composting, or organic cooking. **Loyalty Programs:** Offer discounts or free products to repeat customers. (b) **Promotional Techniques: Sampling:** Distribute free samples at markets or events to introduce products. **Content Marketing:** Share recipes, farming tips, and sustainability stories online. **Local Advertising:** Use local newspapers, flyers, and radio to spread awareness. (c) **Sustainability**

Messaging: Highlight eco-friendly practices (e.g., no chemicals, water conservation, biodiversity promotion) in all communications. (6) **Pricing Strategy**: **Cost-Plus Pricing**: Calculate production costs and add a fair profit margin. **Premium Pricing**: Charge slightly higher for unique, high-quality natural products. **Flexible Pricing**: Offer discounts for bulk purchases or regular subscriptions. (7) **Optimize Post-Harvest Management**: **Storage**: Invest in cold storage or natural preservation methods to maintain product freshness. **Packaging**: Use eco-friendly materials like paper, jute, or biodegradable plastics with farm branding. **Transport**: Ensure quick and efficient delivery systems to reduce spoilage. (8) **Build Community Support**: **Collaborate with Local Farmers**: Create a network to share resources and co-market products. **Participate in Events**: Attend sustainability fairs, food expos, and organic farming conferences to connect with like-minded individuals. **Educate the Community**: Conduct awareness campaigns on the benefits of natural farming. (9) **Monitor and Improve**: **Customer Feedback**: Regularly gather feedback to refine products and services. **Sales Analytics**: Track product popularity, sales trends, and customer preferences. **Expand Gradually**: Start small and expand as demand and resources grow. (10) **Create a Unique Selling Proposition (USP)**: Highlight aspects that differentiate the farm: 100% chemical-free products. Locally sourced seeds or indigenous crop varieties. Transparent farming practices.

18.6 Experimental Design and Layout:

Experimental designs in natural farming research should follow the methods of ecological research. The research methods should be tailored to explore relationships between organisms, their environment, and ecosystem processes. These designs often account for variability, spatial and temporal scales, and interactions among components of ecosystems. Below are some commonly used experimental designs in ecological research:

18.6.1 Observational Studies:

Purpose: Investigate natural variation without manipulating variables. Example: Monitoring species distribution along an environmental gradient (e.g., elevation, temperature). Studying correlations between biodiversity and habitat types. Advantages: Real-world relevance, cost-effective. Limitations: Limited causal inference.

18.6.2 Controlled Experiments:

Purpose: Manipulate one or more variables to study cause–effect relationships. Example: Test the effect of nitrogen addition on plant diversity. Study the impact of predator removal on prey populations. Designs: **Single-Factor**: One variable, multiple levels (e.g., low, medium, high fertilizer). **Multi-Factor**: Interaction between two or more variables (e.g., light and nutrient availability).

18.6.3 Randomized Block Design:

Purpose: Reduces variability by grouping similar experimental units into blocks. Example: Test the effect of grazing intensity on grassland biodiversity, blocking by soil type. Advantages: Controls for spatial or temporal heterogeneity.

18.6.4 Split-Plot Design:

Purpose: Used when treatments vary at two levels. Example: Main plots: Different irrigation levels. Subplots: Different crop species. Advantages: Effective for large-scale ecological experiments.

18.6.5 Factorial Design:

Purpose: Examines interactions between two or more factors. Example: Test the combined effects of temperature and CO_2 concentration on plant growth. Advantages: Identifies interaction effects.

18.6.6 Before-After-Control-Impact (BACI) Design:

Purpose: Evaluates changes due to a specific disturbance or intervention. Example: Assess the effect of dam construction on fish populations before and after the dam is built. Advantages: Accounts for natural variability over time.

18.6.7 Gradient Analysis:

Purpose: Studies responses to continuous changes in an environmental variable. Example: Examine species richness along a moisture gradient. Advantages: Ideal for correlational studies.

18.6.8 Longitudinal Studies:

Purpose: Tracks ecological changes over time. Example: Monitor forest succession over decades after disturbance. Advantages: Provides insights into temporal dynamics.

18.6.9 Cross-Scale Experiments:

Purpose: Combines small-scale manipulations with large-scale observations. Example: Study nutrient cycling at the plot, watershed, and regional scales. Advantages: Links local processes to broader patterns.

18.6.10 Field Mesocosms and Microcosms:

Purpose: Semi-controlled environments to test ecological hypotheses. Example: **Mesocosms:** Test the impact of pollutants in freshwater ecosystems. **Microcosms:** Study microbial responses to temperature changes. Advantages: Controlled yet closer to natural conditions.

18.6.11 Reciprocal Transplant Experiments:

Purpose: Study the effect of local adaptation or environmental conditions. Example: Transplant plants or animals between habitats to observe performance. Advantages: Tests both genetic and environmental influences.

18.6.12 Exclusion Experiments:

Purpose: Remove specific biotic factors (e.g., predators, herbivores) to study effects. Example: Use cages to exclude herbivores and study their impact on plant communities. Advantages: Highlights biotic interactions.

18.6.13 Paired Plot Comparisons:

Purpose: Compare paired sites with and without a treatment. Example: Compare grazed versus ungrazed plots in a grassland ecosystem. Advantages: Controls for site-specific variability.

18.6.14 Landscape-Scale Experiments:

Purpose: Investigate processes across ecosystems. Example: Large-scale manipulation of fire regimes to study their impact on savanna biodiversity. Challenges: Expensive and logistically complex.

18.6.15 Citizen Science and Participatory Research:

Purpose: Engages the public in data collection. Example: Monitoring bird migrations or phenology changes due to climate change. Advantages: Expands spatial and temporal data coverage.

Key Considerations: Replication: Ensures statistical reliability. **Randomization:** Reduces bias. **Scale:** Matches the scale of the experiment to the ecological question. **Metrics:** Select variables like biodiversity, biomass, nutrient cycling, and trophic interactions.

18.7: Data Collection:

18.7.1 Problems of Data Collection in Natural Farming Agroecosystem:

Collecting research data from natural farming agroecosystems presents unique challenges due to the complex and diverse nature of these systems. Here are some key problems:

Complexity of the System: Diverse Interactions: Multilayer polycropping and biodiversity integrity involve multiple interdependent factors, making it challenging to isolate variables for study. **Dynamic Changes:** Natural farming systems are highly dynamic, with seasonal and environmental fluctuations influencing the ecosystem.

Measurement Difficulties: Soil Health Parameters: Assessing soil health in no-till systems is challenging due to variability in organic matter, microbial diversity, and nutrient cycling. **Crop Yields:** Measuring yields accurately in multilayer polycropping systems is difficult because of mixed cropping and non-uniform plant spacing. **Biodiversity Metrics:** Quantifying biodiversity, including flora and fauna, requires intensive observation and expertise in species identification.

Lack of Standardized Methods: Traditional agricultural research often relies on monoculture systems, making it hard to apply standard methods to polycropping or no-till systems. Integrative methods for evaluating ecological, economic, and social outcomes are still under development.

Time and Resource Intensity: Long-term monitoring is required to understand the full impacts of no-till, biodiversity integrity, and other practices on soil health and productivity. Collecting data on natural pest control, pollination, and nutrient cycling involves labor-intensive fieldwork.

Influence of External Factors: External variables like weather, pests, and diseases have a more pronounced impact in natural farming systems, adding noise to research data. The absence of synthetic inputs may lead to site-specific challenges that are hard to generalize.

Farmer Participation and Variability: Farmers may follow diverse practices within the broad principles of natural farming, leading to inconsistent data. Their willingness to share information and cooperate in studies can also be variable.

Technological and Financial Constraints: Advanced tools for soil microbiology, remote sensing, and biodiversity analysis may not be readily available or affordable. Research in natural farming ecosystems often lacks adequate funding compared to conventional farming systems.

Possible Solutions: **Interdisciplinary Approach**: Combine expertise from agronomy, ecology, and social sciences. **Participatory Research**: Engage farmers as co-researchers to ensure practical relevance. **Innovative Technologies**: Use remote sensing, GIS, and AI to analyze large-scale data. **Longitudinal Studies**: Focus on long-term studies to capture systemic changes. **Standardization Efforts**: Develop standardized protocols for natural farming research.

18.7.2 Methods of Sampling Biodiversity (Biodiversity Informatics):

Random Sampling: Individuals are selected purely by chance, minimizing bias and allowing generalization to the entire population. **Stratified Sampling**: The population is divided into distinct subgroups (strata), and samples are drawn from each stratum to ensure representation. **Systematic Sampling:** Samples are taken at regular intervals from a predefined starting point, useful for studying patterns along gradients. **Cluster Sampling:** The population is divided into clusters, with some clusters randomly selected for sampling, often used when individual sampling is impractical. **Convenience Sampling:** Involves selecting easily accessible samples, often used in preliminary studies but may introduce bias. **Purposive Sampling:** Specific individuals are chosen based on predetermined criteria, useful for targeted studies.

18.7.3 Soil Water Measurement:

Soil hydrology measurements are essential for understanding the water movement, retention, and drainage within the soil profile. These measurements are vital for agriculture, natural farming, and land management. Here's an overview of the key parameters and methods used in soil hydrology measurements:

Soil Moisture Content: The amount of water present in the soil, expressed as a percentage of the soil's dry weight. **Measurement Methods: Gravimetric Method:** Weighing soil before and after oven drying. **Time Domain Reflectometry (TDR):** Measures soil's dielectric constant. **Capacitance Sensors:** Measure changes in soil's dielectric properties. **Neutron Probe:** Uses neutron scattering to measure moisture content.

Infiltration Rate: The rate at which water enters the soil surface. **Measurement Methods: Double-Ring Infiltrometer:** Measures water infiltration in a defined area. **Single-Ring Infiltrometer:** Simpler version but can be less accurate. **Rainfall Simulators:** Mimics natural rainfall for large-scale studies.

Hydraulic Conductivity: The soil's ability to transmit water when saturated. **Measurement Methods: Constant-Head Permeameter:** For saturated soil. **Falling-Head Permeameter:**

For less permeable soils. **Field Saturated Hydraulic Conductivity Tests:** In situ measurements.

Field Capacity and Wilting Point: Field Capacity: Maximum water soil can hold after excess water drains. **Wilting Point:** Water content at which plants can no longer extract moisture. **Measurement Methods:** Tension plate or pressure plate apparatus in the laboratory.

Soil Water Retention Curve: Relationship between soil water content and matric potential. **Measurement Methods: Pressure Plate Apparatus:** For controlled lab measurements. **Tensiometers:** Measure soil water potential in the field.

Soil Porosity: The volume of pore spaces in the soil. **Measurement Method:** Calculated from bulk density and particle density.

Runoff and Drainage: Amount of water leaving the soil surface and subsurface. **Measurement Methods: Runoff Plots:** Measure water loss from defined plots. **Lysimeters:** Measure drainage and leaching below the root zone.

Evapotranspiration (ET): The combined water loss from soil and plant surfaces. **Measurement Methods: Eddy Covariance Systems:** Measure atmospheric fluxes. **Lysimeters:** Monitor soil water balance.

18.7.4 Measurement of Soil Organic Matter Content:

Measuring soil organic matter (SOM) is crucial for understanding soil health, fertility, and carbon storage capacity. Various methods are used to quantify SOM, based on its chemical composition and physical properties. Below are the key methods:

Loss-on-Ignition (LOI) Method: Principle: Organic matter is burned off at high temperatures, leaving behind mineral residues. **Procedure:** (1) Dry the soil sample at 105°C to remove moisture. (2) Heat the dried sample in a muffle furnace at 400–550°C. (3)

Calculate the weight loss as the SOM content. **Advantages**: Simple, inexpensive, and widely used. **Limitations**: May overestimate SOM by including other volatile components like carbonates.

Walkley-Black Wet Oxidation Method: **Principle**: Oxidizes organic carbon in the soil using potassium dichromate and sulfuric acid, and measures the unreacted dichromate. **Procedure**: (1) Add potassium dichromate and sulfuric acid to the soil sample. (2) Allow oxidation to occur. (3) Titrate the unreacted dichromate to determine organic carbon. **Advantages**: Quick and reliable for most agricultural soils. **Limitations**: Can underestimate SOM due to incomplete oxidation; uses hazardous chemicals.

Dry Combustion (CHN Analyzer): **Principle**: Burns soil at very high temperatures ($\geq 900\,°C$) to convert organic carbon into CO_2, which is then quantified. **Procedure**: (1) Prepare soil samples by grinding and drying. (2) Analyze in an elemental analyzer (carbon, hydrogen, nitrogen detector). **Advantages**: Highly accurate and precise. **Limitations**: Expensive equipment and requires skilled operation.

Permanganate Oxidation (POXC): **Principle**: Oxidizes labile organic carbon fractions using potassium permanganate, which changes color as it reacts. **Procedure**: (1) Mix soil with potassium permanganate solution. (2) Measure the color change (absorbance) with a spectrophotometer. **Advantages**: Focuses on biologically active carbon, a key indicator of soil health. **Limitations**: Does not measure total organic matter.

Total Organic Carbon (TOC) by Wet Chemistry: **Principle**: Measures carbon content by chemically oxidizing organic matter and quantifying the released carbon dioxide. **Procedure**: Oxidize organic carbon with an acid digestion or combustion. Measure CO_2 using infrared spectroscopy or titration. **Advantages**: Accurate for organic carbon. **Limitations**: Does not differentiate between active and stable organic matter.

Infrared Spectroscopy: Principle: Uses near-infrared (NIR) or mid-infrared (MIR) spectroscopy to analyze organic matter based on its absorption spectrum. **Procedure**: Expose soil to infrared light and record absorption. Use calibration models to estimate SOM. **Advantages**: Rapid, non-destructive, and no chemicals required. **Limitations**: Requires advanced calibration and equipment.

Soil Fractionation Techniques: Principle: Separates organic matter into particulate, mineral-associated, and dissolved organic fractions for targeted analysis. **Procedure**: Use physical (e.g., sieving) and chemical (e.g., density gradient) methods to isolate fractions. Measure organic matter in each fraction. **Advantages**: Provides detailed SOM composition. **Limitations**: Time-consuming and complex.

Remote Sensing Techniques: Principle: Estimates SOM indirectly through soil color and other spectral properties. **Procedure**: Use satellite or drone-based imagery. Apply machine learning models for SOM prediction. **Advantages**: Non-invasive and covers large areas. **Limitations**: Requires ground-truthing and may be less accurate for specific sites.

18.7.5 Soil Health Biomonitoring:

Soil health refers to the health conditions of the living organisms in soil. Natural soil composed of both non-living components such as sand, silt, clay, organic matter, water and air as well as living components such as plants, animals and microorganisms. Due to the presence of living organisms, soil as a whole, behaves like a living organism and it manifests it's own health issues. The health condition of the soil is not static, it changes continuously over time, depending on climatic conditions and other factors. Therefore, continuous monitoring of soil health is required for data collection. Usually the health conditions of the indicator species of plants, animals and microorganisms are visually observed for biomonitoring of soil health. It should be noted that soil health is completely different from soil properties or soil fertility.

Objectives of Biomonitoring Soil Health:

(1) To avoid destructive sampling of soil or living organisms. (2) To maintain the sanctity and integrity of the natural farming agroecosystem. (3) To collect health data from undisturbed living organisms. (4) To collect continuous data. (5) To collect data on interspecific interactions among living organisms in soil. (6) To collect data on the effect of the environment on soil health.

Methods of Biomonitoring Soil Health:

A. Indicator Plants for Biomonitoring of Soil Health:

Indicator plants for biomonitoring of soil health are plant species that reflect the quality, fertility, and health of the soil based on their presence, growth patterns, and physiological conditions. These plants provide valuable insights into soil characteristics such as nutrient availability, pH, compaction, salinity, and contamination. Key Indicator Plants and Their Soil Health Insights

Nutrient Deficiency Indicators: Nitrogen Deficiency: Indicator Plants: Clover (Trifolium spp.), Yellowing grass species. Symptoms: Pale green or yellow leaves, stunted growth.

Phosphorus Deficiency: Indicator Plants: Legumes with poor root nodulation (e.g., soybeans). Symptoms: Purplish discoloration of older leaves. **Potassium Deficiency**:

Indicator Plants: Crops like corn and tomato showing scorched leaf edges. Symptoms: Marginal chlorosis progressing to necrosis.

Soil Acidity (Low pH) Indicators: Indicator Plants: Sorrel (Rumex spp.), Blueberries, Bracken fern (Pteridium aquilinum). Soil

Conditions: Acidic soils (pH < 5.5). Implications: Low availability of calcium and magnesium, potential aluminum toxicity.

Alkaline Soil (High pH) Indicators: Indicator Plants: Saltbush (Atriplex spp.), Pigweed (Amaranthus spp.). Soil Conditions: Alkaline soils (pH > 7.5). Implications: Reduced availability of iron, zinc, and phosphorus.

Compacted Soil Indicators: Indicator Plants: Plantain (Plantago major), Knotweed (Polygonum spp.). Symptoms: Poor root penetration, stunted plant growth. Implications: Reduced aeration and water infiltration.

Saline Soil Indicators: Indicator Plants: Halophytes like Saltgrass (Distichlis spicata) and Seablite (Suaeda spp.). Soil Conditions: High salt content. Implications: Osmotic stress affecting plant growth.

Waterlogged or Poor Drainage Indicators: Indicator Plants: Cattails (Typha spp.), Sedges (Carex spp.), Horsetail (Equisetum spp.). Soil Conditions: Saturated soils with low oxygen availability. Implications: Reduced root health and microbial activity.

Heavy Metal Contamination Indicators: Indicator Plants: Indian Mustard (Brassica juncea) for lead. Sunflowers (Helianthus annuus) for arsenic and cadmium. Symptoms: Stunted growth or tolerance in contaminated soils. Applications: Phytoremediation and pollution detection.

Organic Matter Deficiency Indicators: Indicator Plants: Weedy species like Thistle (Cirsium spp.) and Ragweed (Ambrosia spp.). Soil Conditions: Poor organic matter and microbial activity. Implications: Low fertility and soil structure degradation.

Erosion-Prone Soil Indicators: Indicator Plants: Bare patches with sparse vegetation, invasive grasses. Examples: Cheatgrass (Bromus tectorum), Sandbur (Cenchrus spp.). Implications: Soil compaction and loss of topsoil.

B. Indicator Animals for Biomonitoring of Soil Health:

Indicator animals for biomonitoring of soil health are species whose presence, abundance, and behavior provide valuable insights into soil quality and ecological balance. These organisms often reflect changes in soil structure, fertility, contamination, and biological activity. Key indicator animals and their soil health insights:

Earthworms (Lumbricidae): They Indicate: High organic matter and nutrient-rich soils. Good soil aeration and structure. High earthworm abundance signals healthy, fertile soil. Low or absent populations may indicate compaction, acidity, or contamination. Earthworms improve soil porosity, nutrient cycling, and water infiltration.

Ants (Formicidae): They Indicate: Well-structured, aerated soils. Presence of organic material in the soil. Active ant colonies suggest healthy microbial activity and organic matter. Decline may indicate poor soil health or pesticide exposure. Ants promote soil turnover and nutrient distribution.

Termites (Isoptera): They Indicate: Organic matter presence and decomposition activity.

Presence of termite mounds or tunnels reflects high organic matter and microbial activity. Termites break down plant residues and contribute to nutrient cycling.

Soil Nematodes (Nematoda): They Indicate: Soil nutrient balance and microbial community structure. Beneficial nematodes (bacterivores, fungivores) signal healthy soils. Parasitic nematodes may indicate imbalances or specific crop diseases. They act as bioindicators of soil food web dynamics.

Beetles (Coleoptera): They Indicate: Organic matter and soil fertility. Presence of dung beetles suggests healthy decomposition processes. Decline may indicate soil degradation or pesticide exposure. Beetles contribute to nutrient cycling and soil aeration.

Millipedes and Centipedes (Myriapoda): They Indicate: Soil organic matter and moisture levels. High populations suggest healthy organic matter decomposition. Absence may indicate dryness or compaction. They break down organic material and enrich the soil.

Amphibians (Frogs, Toads): They Indicate: Soil moisture and low contamination levels.

Presence of amphibians reflects good water retention and minimal pollution. Absence may indicate soil contamination, salinity, or waterlogging. Sensitive to changes in soil moisture and pollutants.

Ground–Dwelling Spiders (Araneae): They Indicate: Balanced soil ecosystems and prey availability. High spider abundance indicates a healthy food web and reduced pesticide use. Spiders act as predators, controlling pest populations.

Isopods (Woodlice, Pill Bugs): They Indicate: High organic matter and decomposer activity. Abundant isopods suggest healthy organic matter decomposition. Absence may indicate soil dryness or pollution. They contribute to breaking down organic matter.

Snails and Slugs: They Indicate: Soil moisture levels and organic matter. High populations suggest moist, nutrient-rich soil. Decline may indicate dry or compacted conditions. They aid in the decomposition of plant material.

Methods for Monitoring Soil Indicator Animals: Field Sampling: Use pitfall traps, soil cores, or hand sorting to collect animals for identification. **Quantitative Surveys:** Record the abundance and diversity of indicator species over time. **Behavioral Observation:** Observe feeding, burrowing, or reproductive activities as indicators of soil health. **Environmental Correlation:**

Relate animal populations to soil properties like moisture, pH, and organic content.

C. Indicator Microorganisms for Biomonitoring of Soil Health:

Indicator microorganisms for biomonitoring of soil health are microbial species or groups whose presence, activity, or abundance reflect the condition and quality of the soil ecosystem. These microorganisms play critical roles in nutrient cycling, organic matter decomposition, soil structure maintenance, and pollutant degradation. Key indicator microorganisms and their soil health insights

Nitrogen-Fixing Bacteria: Examples: Rhizobium spp. (in legumes). Azospirillum spp. and Azotobacter spp. (free-living). They Indicate: Soil fertility and nitrogen availability. Nitrogen-fixing bacteria enrich soil nitrogen through symbiotic or free-living processes.

Phosphate-Solubilizing Microorganisms: Examples: Pseudomonas spp. Bacillus spp. Fungi like Aspergillus and Penicillium. They indicate availability of phosphorus in soil and microbial activity. Enhance plant phosphorus uptake by solubilizing inorganic phosphate.

Mycorrhizal Fungi: Examples: Arbuscular mycorrhizal fungi (Glomus spp.). Ectomycorrhizal fungi (e.g., Pisolithus and Laccaria). They Indicate: Soil structure and plant-soil nutrient exchange efficiency. Improve plant uptake of nutrients like phosphorus and water.

Decomposer Microorganisms: Examples: Fungi: Trichoderma spp., Aspergillus spp. Bacteria: Bacillus spp., Actinobacteria. They Indicate: Organic matter decomposition and carbon cycling. Maintain soil organic matter and promote soil fertility.

Denitrifying Bacteria: Examples: Pseudomonas spp., Paracoccus spp., Thiobacillus spp. They Indicate: Soil aeration and nitrogen loss through denitrification. Help assess oxygen levels and nitrate presence in soil.

Cellulose-Degrading Microorganisms: Examples: Fungi: Aspergillus, Trichoderma.

Bacteria: Cellulomonas spp., Bacillus spp. They Indicate: Organic matter decomposition and soil carbon cycling. Break down plant residues and enhance nutrient recycling.

Actinobacteria: Examples: Streptomyces spp., Micromonospora spp. They Indicate:

Soil organic matter turnover and suppression of plant pathogens. Produce antibiotics and improve soil health.

Pathogenic Microorganisms: Examples: Fusarium spp., Verticillium spp. (fungal pathogens). Ralstonia solanacearum (bacterial pathogen). They Indicate: Imbalances in soil microbial communities or poor soil health. Serve as early warning signs of disease outbreaks.

Pollutant-Degrading Microorganisms: Examples: Pseudomonas putida (hydrocarbon degradation). Sphingomonas spp. (pesticide degradation). They Indicate: Soil contamination and bioremediation potential. Break down pollutants and detoxify contaminated soils.

Soil Microbial Biomass: It Indicates: Overall soil biological activity and organic matter content. Measurement: Quantified through soil respiration tests, substrate-induced respiration, or fumigation-extraction methods. Reflects soil fertility and microbial ecosystem health.

Methods for Biomonitoring Microorganisms in Soil: Culture-Based Methods: Isolating and quantifying microorganisms using nutrient-specific media. Example: Counting colony-forming units

(CFUs) of Rhizobium or Pseudomonas. **Molecular Techniques:** DNA-based methods like qPCR, metagenomics, or 16S rRNA sequencing. Example: Detecting specific genes for nitrogen fixation (nifH) or denitrification (nirK). **Soil Enzyme Assays:** Measuring enzyme activities as proxies for microbial functions. Examples: Dehydrogenase activity (general microbial activity). Phosphatase activity (phosphate cycling). **Biomass Carbon and Nitrogen:** Measuring microbial biomass through fumigation-extraction or substrate-induced respiration. **Microbial Respiration:** Measuring CO_2 release to assess microbial activity and organic matter decomposition.

18.7.6 Biomonitoring Plant Health:

Monitoring plant health involves assessing plants for stress, diseases, pests, nutrient deficiencies, and other factors affecting growth. Here are common methods for monitoring plant health:

Visual Inspection: Regular observation of plants for visible signs of stress or disease.

Leaves: Discoloration, wilting, spots, or holes. Stems: Weakness, lesions, or fungal growth. Flowers/Fruits: Poor development, deformities, or rot. Advantages: Low-cost, accessible. Limitations: Subjective; may miss early-stage issues.

Soil Testing: Assessing soil properties to ensure optimal growing conditions. Key Parameters: Nutrient levels (N, P, K). pH balance. Organic matter content. Soil moisture. Tools Used: Soil testing kits, laboratory analysis, soil moisture sensors.

Plant Sap Analysis: Testing plant sap to assess nutrient uptake and stress levels.

What It Detects: Nutritional imbalances. Disease and pest stress. Advantages: Provides real-time insights.

Remote Sensing: Using drones, satellites, or sensors to monitor plant health over large areas. Techniques: **Multispectral Imaging**: Detects chlorophyll levels and plant vigor.

Thermal Imaging: Identifies water stress. Advantages: Non-invasive, suitable for large-scale farms.

Internet of Things (IoT) Sensors: Real-time monitoring using sensors in the field.

Key Parameters Measured: Soil moisture and temperature. Air humidity and temperature. Nutrient levels. Tools: Smart farming systems like soil probes and weather stations.

Biochemical Testing: Measuring biochemical markers in plants to assess health. Tests:

Chlorophyll fluorescence for photosynthetic efficiency. Enzyme activity for stress response.

Laboratory Testing: Detailed analysis of plant tissues for pathogens, toxins, or deficiencies. Methods: PCR for pathogen detection. Microscopy for pest and disease identification.

Digital Imaging and AI Tools: Using image recognition software or apps to diagnose plant health issues. How It Works: Analyzes leaf color, shape, and texture to detect abnormalities. Examples: Smartphone apps for farmers and gardeners.

Pest and Disease Monitoring: Tracking the presence and activity of pests and diseases.

Methods: Sticky traps for flying insects. Scouting for pest damage or disease symptoms.

Growth Monitoring: Measuring plant growth parameters as indicators of health. Key Parameters: Height and girth of stems. Leaf area and biomass.

Use of Bioindicators: Monitoring specific plant species or parts for environmental stress markers. Example: Early leaf discoloration in indicator species signaling nutrient deficiency.

Integrated Monitoring Systems: Combining multiple methods (e.g., IoT sensors with remote sensing and visual inspections) ensures a comprehensive understanding of plant health.

Biomonitoring Environment Health:

Biomonitoring Environmental Health involves the use of biological organisms or biological samples to assess the quality of the environment and detect pollution or other environmental changes. Here are the primary methods:

Bioindicator Species Monitoring: Observing species whose presence, abundance, or health reflects environmental conditions. Examples: Lichens and Mosses: Sensitive to air quality and heavy metal pollution. Aquatic Insects: Indicators of water quality (e.g., mayflies for clean water). Advantages: Simple and cost-effective. Limitations: May require expert identification.

Biomarker Analysis: Measuring biochemical, cellular, or physiological changes in organisms exposed to pollutants. Key Biomarkers: Enzyme activity (e.g., acetylcholinesterase for pesticide exposure). DNA damage in cells. Stress proteins or hormones. Advantages: Early detection of pollution impacts.

Passive Sampling Using Organisms: Using organisms like mussels, fish, or plants to accumulate pollutants over time. Examples: Mussels Watch Programs: Monitor heavy metals and hydrocarbons in coastal waters. Plant Monitoring: Accumulation of airborne pollutants like sulfur or nitrogen. Advantages: Provides integrated exposure over time.

Population and Community Assessments: Evaluating biodiversity, abundance, and species composition in ecosystems. Methods: Monitoring population trends of sensitive species. Assessing shifts in community structure (e.g., dominance of pollution-tolerant species). Example: Using benthic macroinvertebrates to assess river health.

Genetic Biomonitoring: Analyzing genetic changes in organisms to detect environmental stress. Techniques: **DNA Barcoding:** Identifying changes in species composition.

Microsatellite Analysis: Detecting genetic diversity loss due to pollution. Advantages: Highly precise and informative.

Bioassays: Exposing test organisms to environmental samples to measure toxicity. Examples: **Daphnia Toxicity Test:** Monitoring water quality using water fleas. **Algal Growth Tests:** Assessing nutrient levels and toxic substances. Advantages: Provides quantitative toxicity data.

Remote Sensing of Biological Indicators: Using satellite imagery or drones to monitor vegetation, algal blooms, and habitat conditions. Applications: Identifying areas affected by deforestation or eutrophication.

Monitoring coral bleaching events.

Physiological and Behavioral Monitoring:

Observing changes in organism behavior or physiology in response to pollutants. Examples: Changes in fish swimming patterns due to water contamination. Leaf chlorosis in plants exposed to air pollutants.

Ecosystem Process Monitoring: Measuring changes in processes like decomposition, nutrient cycling, or primary production. Methods: Litter decomposition rates in forests.

Algal productivity in water bodies:

Integrated Biomonitoring Programs: Combining multiple methods to provide a comprehensive assessment. Examples: **Air Quality Monitoring Programs:** Using lichens, dust analysis, and human health data. **Aquatic Ecosystem Assessments:** Using fish, macroinvertebrates, and chemical sampling.

Applications of Environmental Biomonitoring: Air Quality: Lichens, mosses, and particulate matter analysis. **Water Quality:** Aquatic invertebrates, fish bioassays, and algal blooms. **Soil Health:** Earthworms and microbial activity indicators. **Marine Health:** Coral monitoring and shellfish pollutant accumulation.

18.7.7 Biomonitoring Natural Control:

Biomonitoring natural control of pests involves assessing the activity and effectiveness of natural enemies, such as predators, parasitoids, and pathogens, in regulating pest populations.

Direct Observation and Counting: Physically observing natural enemies and pest populations in the field. Methods: **Visual Surveys:** Counting predators (e.g., ladybirds, spiders) and parasitoids on crops. **Sampling Pests:** Collecting pest individuals and assessing parasitism or predation rates. Low cost and straightforward. Limitations: Time-consuming and may miss nocturnal or cryptic species.

Trapping Techniques: Using traps to monitor the presence and abundance of natural enemies. Types of Traps: **Sweep Net Sampling:** Use a sweep net to capture insects from crops and surrounding vegetation. Identify and count natural enemies like ladybirds, lacewings, and predatory beetles. **Pitfall Traps:** For ground-dwelling predators like beetles and spiders. **Sticky Traps:** To capture flying parasitoids or predatory insects. **Light Traps:** To monitor nocturnal natural enemies and pests. **Pheromone Traps:**

To monitor pests of crop plants. Advantages: Provides continuous monitoring data.

Biodiversity Assessment: Species Richness and Abundance: Assess the diversity and abundance of natural enemies in the crop ecosystem. High biodiversity often correlates with better pest control. **Indicator Species:** Identify key indicator species that reflect the health of the pest control ecosystem, such as specific predatory beetles or parasitoid wasps.

Assessing Parasitism and Predation: Parasitism Rate: Collect pest samples (e.g., caterpillars or aphids) and rear them in controlled conditions to identify emerging parasitoids. Calculate parasitism rates to measure the effectiveness of parasitoids like Trichogramma wasps or tachinid flies. **Predation Rate:** Use sentinel prey (e.g., pest eggs or larvae) placed in the field to assess predation levels. Record the proportion of prey removed or consumed over a set time.

Molecular Methods: Using DNA-based techniques to detect natural enemies or predation events. Methods: **DNA Barcoding:** Identifies natural enemies or pests from collected samples. **Gut Content Analysis:** Detects pest DNA in the gut of predators to confirm predation. Advantages: Highly accurate and can reveal hidden interactions. **Environmental DNA (eDNA):** Collect and analyze soil or water samples for DNA traces of natural enemies and pests. Provides insights into the presence and activity of beneficial organisms.

Mark-Release-Recapture (MRR): Marking natural enemies and tracking their movement and activity in the field. Methods: Using dyes, powders, or radio tags to mark individuals.

Recapturing marked individuals to assess dispersal and population density. Applications: Monitoring the effectiveness of released biocontrol agents.

Crop Damage Assessment: Damage Surveys: Examine crop plants for pest damage and correlate with the presence of natural enemies. Reduced damage in areas with high natural enemy activity indicates effective control. **Yield Analysis:** Compare yields in areas with active biomonitoring of natural enemies to those with chemical pest control.

Exclusion Experiments: Comparing pest populations or damage in areas where natural enemies are excluded versus accessible. Methods: Using cages or barriers to prevent access by predators or parasitoids. Monitoring pest population growth in protected and unprotected areas. Advantages: Directly quantifies the impact of natural enemies.

Behavioral Observation: Monitoring the foraging, hunting, or parasitizing behavior of natural enemies. Examples: Observing wasps attacking caterpillars. Tracking predatory beetles feeding on aphids. Advantages: Provides real-time insights into ecological interactions.

Artificial Refugia and Habitat Manipulation: Monitoring in Refugia: Install artificial habitats (e.g., insect hotels or hedgerows) to attract and monitor natural enemies. Track the colonization and pest control activity in these habitats. **Floral Resources:** Plant flowering strips to attract pollinators and parasitoids. Monitor their impact on pest populations in adjacent crops.

Pathogen Monitoring in Pests: Assessing the prevalence of microbial pathogens (e.g., fungi, bacteria, viruses) that naturally control pest populations. Methods: Isolating pathogens from pest cadavers. Molecular diagnostics to identify pathogen species.

Applications: Evaluating the effectiveness of microbial biocontrol:

Environmental Monitoring for Natural Enemy Activity: Assessing environmental factors that influence the activity of natural enemies. Parameters Monitored: Temperature, humidity, and vegetation cover. Availability of alternative prey or floral resources. Tools: Weather stations, remote sensing, and habitat mapping.

Remote Sensing and Imaging: Drone Surveys: Use drones equipped with high-resolution cameras to monitor pest and natural enemy activity. Detect patterns of pest outbreaks and predator activity over large areas. **Thermal Imaging:** Identify hotspots of insect activity based on temperature variations, which may indicate pest control zones.

Population Modeling and Data Analysis: Using statistical models to predict pest suppression based on natural enemy dynamics. Data Sources: Field surveys, trapping data, and environmental conditions.

Advantages: Helps plan interventions and optimize biocontrol strategies.

Farmer Participation: Citizen Science: Engage farmers in recording sightings of natural enemies and pests. Use mobile apps or logbooks to gather data from multiple fields. **Pheromone Traps:** Set up pheromone traps for pests and monitor the impact of natural enemies on trap captures.

Long-term Monitoring Programs: Establish permanent monitoring plots to track natural pest control trends over multiple seasons. Collect data on climatic conditions, crop rotation, and pesticide use to correlate with natural pest control effectiveness.

18.7.8 Biomonitoring Pollination:

Pollinator Diversity and Abundance: Measure the variety and number of pollinators visiting crops. **Methods: Field Surveys:** Observing and counting pollinators in a defined area during bloom. **Trapping Techniques:** Using pan traps, netting, or bait to capture pollinators for identification. **Molecular Techniques:** DNA barcoding for accurate species identification.

Pollination Efficiency: Evaluate the effectiveness of specific pollinators in transferring pollen. **Methods: Pollen Load Analysis:** Assessing the quantity and type of pollen carried by pollinators. **Single-Visit Fruit Set:** Measuring fruit or seed set from a single visit by a pollinator.

Flower Visitation Rates: Understand pollinator activity and behavior. **Methods:** Recording the number of visits per flower per unit of time. Using video or motion-sensitive cameras for observation.

Crop Yield and Quality Assessment: Correlate pollination levels with crop productivity and quality. **Methods:** Comparing yield in open-pollinated vs. pollinator-excluded areas (e.g., bagged flowers). Measuring parameters like fruit weight, seed set, or oil content.

Environmental and Habitat Factors: Assess how habitat and landscape features influence pollinator activity. **Methods:** Analyzing surrounding vegetation diversity. Monitoring weather conditions (temperature, wind speed, humidity).

Technologies in Pollination Biomonitoring:

Remote Sensing and Imaging: Drones and multispectral cameras for assessing floral density and pollinator activity from above. **Radio-Frequency Identification (RFID) Tags:** Tracking pollinator movement and foraging patterns in real time. **Automated Pollinator Monitoring Systems:** Devices with cameras and AI algorithms to identify and count pollinators. **Citizen Science Platforms:** Engaging the public to monitor pollinator activity using mobile apps.

18.7.9 Measurement of Carbon Sequestration:

Carbon sequestration refers to the process of capturing and storing atmospheric carbon dioxide (CO_2) to mitigate climate change. The measurement of carbon sequestration involves assessing the amount of carbon captured and stored in various systems, such as soil, vegetation, and oceans. Below are the primary methods:

Soil Carbon Sequestration: Soil Sampling and Analysis: Collect soil samples at different depths. Analyze organic carbon content using methods like dry combustion or wet oxidation. **Remote Sensing:** Use satellite imagery and ground-based sensors to estimate soil carbon content over large areas. **Modeling Approaches:** Use models like Century, RothC, or DNDC to simulate soil organic carbon changes based on management practices.

Carbon Sequestration in Agroforestry Systems: Tree and Crop Biomass Measurements: Apply similar methods as in forest systems. **Soil Organic Carbon:** Measure changes in soil carbon under agroforestry practices. **Ecosystem Modeling:** Use models like Agroforestry Carbon Calculator (AFCC) to estimate carbon storage.

Carbon Sequestration Verification: Life Cycle Assessment (LCA): Evaluate the net carbon sequestration considering all emissions and removals. **Carbon Credit Auditing:** Use third-party verifications to confirm carbon sequestration claims for trading or certification.

18.7.10 Valuation of Ecosystem Services:

Valuing ecosystem services involves assigning economic, social, or intrinsic value to the benefits ecosystems provide to humans. These services are categorized into provisioning, regulating, cultural, and supporting services. The methods of valuation are typically grouped into market-based, non-market-based, and hybrid approaches. Below is an overview of the main methods:

Market-Based Valuation:

These methods use market prices or observable economic behaviors to value ecosystem services. **Market Price Method:** Description:

Directly uses market prices for goods or services (e.g., timber, fish, water). Example: Valuing timber extracted from a forest at the current market price per cubic meter. **Productivity Method (Production Function Approach)**: Description: Estimates the contribution of an ecosystem service to the production of a marketable good. Example: Valuing pollination services based on the increased agricultural yield they support. **Replacement Cost Method**: Description: Estimates the cost of replacing an ecosystem service with man-made alternatives. Example: Valuing water filtration by wetlands based on the cost of building a water treatment plant. **Avoided Cost Method**: Description: Measures the costs avoided due to the presence of ecosystem services. Example: Valuing coastal mangroves for storm protection by calculating the cost of damages avoided.

Non-Market-Based Valuation:

These methods assess the value of services not traded in markets, often using proxies. **Contingent Valuation (Willingness to Pay or Accept)**: Description: Uses surveys to ask people their willingness to pay (WTP) to preserve or enhance an ecosystem service, or willingness to accept compensation for its loss. Example: Assessing public willingness to pay for conserving a national park. **Hedonic Pricing**: Description: Derives value from differences in market prices of related goods, often real estate. Example: Higher property values near clean lakes or forests reflect the value of these ecosystems. **Travel Cost Method**: Description: Estimates value based on how much people spend to visit a site. Example: Valuing recreational services provided by a forest or beach based on visitor expenses (travel, accommodation, etc.). **Benefit Transfer Method**: Description: Uses existing valuation estimates from similar ecosystems elsewhere. Example: Applying the value of wetland services in one region to a similar wetland in another region.

Hybrid Methods:

These integrate multiple data sources or combine market and non-market techniques. **Cost–Benefit Analysis (CBA):** Description: Compares the costs and benefits (both monetary and non-monetary) of a project or policy affecting ecosystems. Example: Valuing ecosystem restoration projects by assessing long-term benefits like improved water quality and biodiversity. **Ecosystem Services Index:** Description: Assigns a composite value to multiple services provided by an ecosystem. Example: Combining the value of carbon sequestration, water purification, and recreational benefits of a forest.

Intrinsic and Cultural Valuation:

These methods assess non-economic values tied to human perceptions, ethics, or spirituality. **Deliberative Valuation:** Description: Involves group discussions to assess collective preferences for ecosystem services. Example: Valuing cultural landscapes by community consensus. **Participatory Mapping:** Description: Uses local knowledge to map and assess the value of ecosystem services. Example: Mapping spiritual or heritage sites in a forest. **Multi–Criteria Decision Analysis (MCDA):** Description: Evaluates ecosystem services based on multiple, often non-monetary, criteria. Example: Ranking forest conservation projects based on biodiversity, carbon storage, and recreational values.

Biophysical Valuation:

Focuses on quantifying ecosystem services in physical terms, which can later be monetized. **Carbon Pricing:** Description: Assigns a value to carbon sequestration based on carbon market prices. Example: Valuing forest carbon sequestration at the prevailing carbon credit price. **Energy Analysis:** Description: Measures the energy saved or provided by ecosystem functions. Example: Valuing windbreaks that reduce heating and cooling costs.

18.7.11 Methods of Ecological Economics:

Ecological economics employs various methods to analyze the complex interactions between economies and ecosystems, emphasizing sustainability, equity, and resilience. These methods are interdisciplinary, drawing from economics, ecology, sociology, and systems science. Below are the key methods used in ecological economics:

Material Flow Analysis (MFA): Tracks the flow of materials (e.g., energy, water, raw materials) through the economy.

Quantifies resource extraction, processing, consumption, and waste generation. Helps identify inefficiencies and opportunities for reducing resource use.

Ecological Footprint Analysis: Measures the amount of biologically productive land and water required to support a population's consumption and absorb its waste. Compares the ecological footprint with Earth's biocapacity to determine sustainability. Highlights overshoot, where human demand exceeds planetary limits.

Energy and Exergy Analysis: Energy Analysis: Evaluates the energy inputs and outputs of economic processes.

Exergy Analysis: Assesses the quality and usability of energy and materials, focusing on entropy and degradation. These methods reveal the energetic efficiency and sustainability of economic activities.

Cost-Benefit Analysis (CBA) with Environmental Valuation: Incorporates environmental and social costs/benefits into traditional CBA. Uses methods like contingent valuation, hedonic pricing, and travel cost method to assign monetary values to non-market goods (e.g., clean air, biodiversity).

Life Cycle Assessment (LCA): Evaluates the environmental impacts of products or processes across their entire life cycle, from resource extraction to disposal. Helps identify stages with the highest environmental impact and improve sustainability.

Input-Output Analysis (IOA): Analyzes the interdependencies between different sectors of the economy. Extended in ecological economics to include environmental factors like energy use, emissions, and resource flows. Helps quantify the environmental impact of economic activities.

Systems Dynamics Modeling: Simulates the interactions and feedback loops between economic, social, and ecological systems. Used to predict the outcomes of policies and identify tipping points or thresholds (e.g., planetary boundaries). Tools like STELLA and Vensim are often used.

Multi-Criteria Decision Analysis (MCDA): Evaluates multiple, often conflicting, criteria to support decision-making. Incorporates environmental, social, and economic factors without reducing them to a single monetary value. Used in stakeholder-driven decision-making processes.

Scenario Analysis: Explores possible future outcomes based on different assumptions about technological, economic, and policy changes. Helps policymakers evaluate the long-term implications of their choices under uncertainty.

Ecosystem Service Valuation: Quantifies the economic value of ecosystem services (e.g., pollination, carbon sequestration). Aims to integrate the value of natural capital into decision-making.

Participatory and Stakeholder Engagement Methods: Engages stakeholders (communities, businesses, policymakers) to incorporate diverse perspectives and values. Methods include focus groups, deliberative workshops, and participatory mapping.

Dynamic Optimization and Game Theory:

Models strategic interactions between agents (e.g., governments, firms) to study resource allocation and cooperation. Explores solutions to common-pool resource problems and environmental conflicts.

Emergy Analysis: Measures the total energy required to produce a good or service, including direct and indirect energy inputs. Expresses all resources in terms of a common energy unit (e.g., solar emergy joules).

Planetary Boundaries Framework: Assesses human impacts on key Earth system processes (e.g., climate change, biodiversity). Identifies thresholds that should not be crossed to maintain a stable environment.

Social and Institutional Analysis: Examines the social and institutional drivers of environmental degradation and sustainability. Explores cultural, ethical, and behavioral dimensions influencing resource use.

GIS and Spatial Analysis: Uses Geographic Information Systems (GIS) to analyze spatial patterns of resource use, biodiversity, and land-use change. Helps in planning and managing ecological-economic systems.

18.8 Data Analysis:

Analyzing research data on natural farming, which involves complex agroecosystems, requires a mix of statistical and mathematical methods to account for the variability, interdependence, and nonlinear dynamics of such systems. Below is an overview of applicable methods:

18.8.1 Statistical Analysis Methods:

Descriptive Statistics: Purpose: Summarize and describe the main features of data. Examples: Mean, median, variance, and standard deviation for yield, soil organic matter, biodiversity indices, etc. Frequency distribution of crop yields in polyculture systems.

Inferential Statistics: Hypothesis Testing: t-tests/ANOVA: Compare means between treatments (e.g., polyculture vs. monoculture).

Chi-square Tests: Analyze categorical data, such as pest incidence across different farming practices.

Regression Analysis: Linear Regression: Explore relationships between variables (e.g., soil nitrogen and crop yield). **Nonlinear Regression:** Model complex relationships (e.g., biodiversity vs. productivity).

Multivariate Analysis: Principal Component Analysis (PCA): Reduce dimensionality and identify key variables influencing ecosystem dynamics. Example: Determining key factors affecting soil health in polyculture systems.

Cluster Analysis: Group similar ecological plots or farming practices based on multiple variables. Example: Grouping farms based on biodiversity or yield patterns.

Canonical Correspondence Analysis (CCA): Explore relationships between species composition and environmental variables.

Mixed–Effect Models: Purpose: Handle hierarchical or nested data (e.g., plots within farms). Example: Assess crop yield variability while accounting for differences in soil types across farms.

Time Series Analysis: Purpose: Analyze changes over time in agroecosystems. Methods: Seasonal decomposition to identify trends in crop yield or pest dynamics. ARIMA models for forecasting long-term productivity trends.

Spatial Analysis: Geostatistics: Analyze spatial patterns in soil fertility, crop yields, or pest distribution. Methods: **Kriging or Inverse Distance Weighting (IDW)** for mapping soil properties. **Spatial autocorrelation** (e.g., Moran's I) to study spatial dependence.

Structural Equation Modeling (SEM): Purpose: Explore causal relationships among multiple variables. Example: Assess the impact of farming practices on soil health, biodiversity, and yield simultaneously.

18.8.2 Mathematical Analysis Methods:

Systems Dynamics Modeling: Purpose: Model complex interactions and feedback loops in agroecosystems. Example: Simulate the impact of intercropping on pest control, nutrient cycling, and yield.

Differential Equations: Purpose: Model changes in ecological or agronomic variables over time or space. Example: Use Lotka-Volterra equations to model predator-prey (pest-natural enemy) dynamics in polycultures.

Optimization Techniques: Linear Programming: Optimize resource allocation (e.g., land, water, labor) for maximum yield and sustainability. **Nonlinear Optimization:**

Address complex constraints in polyculture systems (e.g., optimal crop combinations).

Ecological Network Analysis: Purpose: Understand energy, nutrient, or material flows in agroecosystems. Example: Analyze trophic interactions or nutrient cycles in a polyculture system.

Stability and Resilience Analysis: Lyapunov Functions: Assess the stability of agroecosystems under disturbances. **Resilience Indices:** Quantify the ability of a polyculture system to recover after shocks (e.g., drought, pest outbreaks).

Bayesian Modeling: Purpose: Incorporate uncertainty and prior knowledge in model predictions. Example: Predict pest outbreak probabilities under different intercropping strategies.

Agent–Based Modeling: Purpose: Simulate individual–based interactions (e.g., farmer decisions, pest movements). Example: Study the impact of farmer behavior on the adoption of ecological practices.

18.8.3 Integrated Approaches:

Given the complexity of natural farming systems, it is often necessary to combine statistical and mathematical methods: **Coupled Models**: Combine time series analysis (statistical) with systems dynamics modeling (mathematical). **Decision Support Tools**: Use optimization techniques alongside regression models to guide sustainable farming decisions. **Data–Driven and Mechanistic Models**: Use machine learning for pattern detection and mechanistic models for hypothesis testing.

18.8.4 Software and Tools:

Statistical Analysis: R, SPSS, SAS, Python (e.g., NumPy, SciPy, Statsmodels). **Mathematical Modeling**: MATLAB, Mathematica, Python (e.g., SymPy, SciPy). **Spatial and Multivariate Analysis**: QGIS, ArcGIS, R (vegan, ade4 packages). **Simulation and Systems Modeling**: STELLA, AnyLogic, Vensim, NetLogo.

18.9 Publication of Results:

Publishing research results on natural farming, particularly for highly complex agroecosystems, can be challenging due to the uniqueness and interdisciplinary nature of the field. The following are key issues related to the publication process:

18.9.1 Scientific Challenges:

Complexity of Agroecosystems: Issue: Natural farming involves complex, interconnected processes (e.g., soil health, biodiversity, water cycles) that are difficult to isolate and quantify. Impact: Reviewers and journals may question the robustness of findings due to the inherent variability and lack of clear cause-effect relationships. **Lack of Standardized Methodologies**: Issue: No universal protocols exist for measuring or evaluating natural farming outcomes (e.g., yield, ecosystem services, or resilience). Impact: Results may appear inconsistent or incomparable to studies using conventional farming metrics. **Long-Term Results and Data Gaps**: Issue: Natural farming benefits, such as improved soil health or carbon sequestration, often take years to materialize. Impact: Short-term studies may miss key outcomes, leading to incomplete or inconclusive results. **Interdisciplinary Nature**: Issue: Research on natural farming often spans agronomy, ecology, economics, and social sciences. Impact: Finding journals that accommodate interdisciplinary studies can be difficult.

18.9.2 Data and Statistical Issues:

Variability in Results: Issue: High variability in natural farming outcomes due to factors like location, soil type, and crop diversity. Impact: Significant variability may reduce statistical power and lead to skepticism about the findings. **Challenges in Statistical Modeling**: Issue: Conventional statistical tools may not adequately capture non-linear interactions and feedback loops in agroecosystems. Impact: Misinterpretation of results or rejection by reviewers unfamiliar with advanced modeling approaches. **Lack of Big Data**: Issue: Studies on natural farming often have small sample sizes due to resource constraints. Impact: Limited generalizability and difficulty in drawing robust conclusions.

18.9.3 Perception and Acceptance Issues:

Bias Toward Conventional Agriculture: Issue: Many journals and reviewers are more familiar with conventional farming practices and metrics (e.g., yield maximization). Impact: Results emphasizing ecosystem services or sustainability may face skepticism or be undervalued. **Resistance to Non-Traditional Practices**: Issue: Practices like no-till, no-GMO, and no off-farm inputs challenge mainstream agricultural paradigms. Impact: Journals may perceive the research as niche or lacking practical relevance. **Overemphasis on Yield**: Issue: Yield is often the primary metric for assessing agricultural systems. Impact: Journals may dismiss studies that focus on broader ecological or social benefits of natural farming.

18.9.4 Ethical and Practical Issues:

Proprietary Practices and Knowledge: Issue: Natural farming often integrates traditional and indigenous knowledge, which may be undervalued or misappropriated. Impact: Ethical concerns over intellectual property rights and acknowledgment. **Funding Bias**: Issue: Many agricultural studies are funded by entities favoring conventional systems (e.g., agrochemical companies). Impact: Journals may scrutinize studies on natural farming for perceived lack of neutrality or funding support. **Difficulty in Peer Review**: Issue: Few reviewers have expertise in complex agroecosystems or natural farming practices. Impact: Longer review times and inconsistent feedback.

18.9.5 Communication Issues:

Terminology Differences: Issue: Terms like "no-till" or "natural farming" may not be universally understood or accepted. Impact: Misinterpretation of methods and results by readers or reviewers. **Difficulty in Presenting Complex Results**: Issue: Results often involve multidimensional outcomes (e.g., trade-offs between yield, biodiversity, and soil health). Impact: Challenges in clearly communicating findings through conventional formats like tables or graphs. **Lack of Accessible Platforms**: Issue: Few high-impact journals focus on agroecology or sustainable agriculture. Impact: Difficulty in reaching broader audiences, including policymakers and practitioners.

18.9.6 Strategies to Overcome Challenges:

Use Interdisciplinary Journals: Target journals that specialize in agroecology, sustainability, or ecological economics (e.g., Agriculture, Ecosystems & Environment, Ecological Applications). **Adopt Advanced Analytical Tools**: Use systems modeling, network analysis, or spatial statistics to capture complexity and validate findings. **Emphasize Ecosystem Services**: Highlight benefits beyond yield, such as carbon sequestration, biodiversity conservation, and water regulation. **Collaborate Across Disciplines**: Partner with ecologists, economists, and social scientists to produce holistic studies. **Transparent Data Sharing**: Publish raw data in open repositories to increase credibility and reproducibility. **Improve Communication**: Use visualizations, case studies, and simplified language to make findings more accessible to non-specialist audiences. **Focus on Long-Term Studies**: Secure funding for longitudinal research to demonstrate sustained benefits of natural farming.

18.10 Sources of Fund for Research on Natural Farming:

Securing funds for research on natural farming can involve exploring multiple funding sources, including government grants, private organizations, international institutions, and collaborations. Below is a categorized list of potential funding sources:

18.10.1 Government Funding:

India–Specific Programs: ICAR (Indian Council of Agricultural Research): Offers grants for agricultural research through schemes like the **National Agricultural Innovation Project (NAIP)**. **Rashtriya Krishi Vikas Yojana (RKVY)**: Funds projects on sustainable and organic farming practices. **NABARD** (National Bank for Agriculture and Rural Development): Provides funding for research and pilot projects related to natural farming and agroecology. **State Government Schemes**: States like Andhra Pradesh, Sikkim, and Uttarakhand have specific funding for promoting natural and organic farming.

18.10.2 International Organizations:

FAO (Food and Agriculture Organization): Supports projects focusing on sustainable farming practices and agroecology. **UNDP** (United Nations Development Programme): Funds initiatives on biodiversity, sustainable agriculture, and climate–resilient farming. **World Bank** and Global Environment Facility (GEF): Provide grants for natural resource management and sustainable agriculture. **International Fund for Agricultural Development (IFAD)**: Funds research on improving rural livelihoods through sustainable farming.

18.10.3 Private Foundations and Trusts:

Tata Trusts: Offers grants for research on sustainable agriculture in India. **Azim Premji Foundation**: Focuses on rural development, including sustainable farming practices.

Rockefeller Foundation: Funds projects on agroecology and food systems. **Bill & Melinda Gates Foundation**: Supports agricultural innovations, including sustainable practices. **Ford Foundation**: Provides funding for ecological and sustainable development projects.

18.10.4 Research Institutions and Universities:

National Agricultural Universities: Offer funding for faculty and student research in sustainable farming. **International Agricultural Research Centers:** Centers like ICRISAT (International Crops Research Institute for the Semi-Arid Tropics) often fund collaborative research. **Corporate-Academia Partnerships:** Companies interested in sustainable sourcing may fund natural farming research through universities.

18.10.5 Corporate Social Responsibility (CSR) Initiatives:

Indian companies are mandated to allocate 2% of profits toward CSR activities.

Example: **ITC Limited:** Funds sustainable agriculture projects. **Mahindra & Mahindra:** Supports rural development and farming practices through CSR.

18.10.6 Non-Governmental Organizations (NGOs):

ActionAid India: Supports agroecology and natural farming projects. **WWF-India:** Funds research on biodiversity-friendly farming practices. **Centre for Science and Environment (CSE):** Supports projects on sustainable agriculture.

18.10.7 Crowdfunding and Philanthropy:

Platforms like Ketto, Milaap, and GoFundMe can be used to gather public funding for research projects. Collaboration with philanthropists interested in sustainability can also provide resources.

18.10.8 Industry Partnerships:

Collaborate with companies in: Organic fertilizer and bio-pesticide sectors. Agri-tech startups focused on sustainable solutions.

18.10.9 International Research Grants:

Horizon Europe (EU): Offers grants for sustainable agriculture research. **USAID** (United States Agency for International Development): Funds projects on rural development and sustainable farming. **CGIAR** (Consultative Group on International Agricultural Research): Supports research on agroecology and natural farming.

18.10.10 Self-Funding and Revenue Models:

Publications and Consultancy: Publishing of books can generate funds. Offering consultancy services in natural farming can supplement funding. **Private Companies:** Revenue from own companies can be reinvested into research initiatives.

DIFFERENCE BETWEEN CONVENTIONAL AGRICULTURE, ORGANIC FARMING AND NATURAL FARMING

Parameter	Conventional Agriculture	Organic Farming	Natural Farming
Innovator	Non-farmer Scientists	Non-farmer Agronomists	Indigenous People, Farmers
Regulator	Industries (seed, input, machine), Governments	IFOAM	Farmers themselves
Certification	Not applicable	Organic Certification by IFOAM	Not necessary (Premium Product)
Motive	Maximising profit for Industries, Food Security of Citizens	Maximising profit of Industries (organic Inputs), and Organic Certifying Agencies	Maximising profit of Farmers, Nutritional Security, Health and Immunity of Consumers, Mitigation of Climate Change, Biodiversity Loss, Pollution
Plant Breeding Objectives	Maximise Yield, Maximise Input Consumption	Not Applicable	Maximise Product Quality, Maximise Yield under no-till, no-input, no-irrigation conditions, Maximising Product Price
Seed	HYV, Hybrid, GM Seed	Organic Seed (Landrace, Heirloom Seed)	Natural OP Seed

Parameter	Conventional Agriculture	Organic Farming	Natural Farming
Tillage	Deep Tillage (Clean Cultivation)	Conservation Tillage	No-Till
Cropping System	Monoculture	Monoculture	Polyculture (Multilayer Polycropping)
Irrigation	Essential	Essential	Not necessary (Rainfed, Ecosystem Services)
Plant Nutrition	Synthetic Chemical Inputs (fertilisers, micronutrients, plant growth regulators)	Positive List of Inputs (Inorganic, Organic, Biological), Crop Rotation, Negative List of Chemical Inputs	Ecosystem Services (biomass decom position, mineralisation, mutualism, nutrient cycles), Probiotics
Crop Protection	Synthetic Chemical Inputs (herbicides, pesticides, fungicides, antibiotics)	Integrated Pest Management (IPM) (physical, mechanical, cultural control, organic and biopesticides, insect traps, crop rotation)	Ecosystem Services (Natural Control of insect pests, diseases and weeds)
Pollination	Seasonal Beekeeping	Not Applicable	Pollinator Sanctuary, Permanent Beekeeping (Native Bees)
Livestock	Not Applicable	Not Applicable	Integrated Farming

Parameter	Conventional Agriculture	Organic Farming	Natural Farming
Cost of Cultivation	Very High	Very High	Near Zero
Profit Margin	Negative (Need Subsidy)	Low (Need Organic Certification)	Highest
Risk Probability	100%	100%	[(1/number of crops) X 100]%
Trade -Offs	Deforestation, Habitat Loss, Soil Erosion, Soil Degradation, Desertification of Land, Genetic Erosion, Biodiversity loss, Climate Change, Infestation of insect pests, diseases and weeds, development of resistance against pesticides, malnutrition, hunger, poverty, human diseases, indebtedness, farmers suicide	Low Demand, Limited Market Size, High Input Cost, Low Efficacy of Organic Inputs, Low Profit Margin	Unwanted Biodiversity (e.g. snakes), Conflict of Interest with Industries, Complex Agroeco system, Steep Learning Curve

Parameter	Conventional Agriculture	Organic Farming	Natural Farming
Research & Development	By Institutions, Non-Farmer Scientists	Not Applicable	By Individual Frmers
IPR Protection	Seed Act, Patent	Not Applicable	The Protection of Plant Varieties and Farmers Rights Act, 2001

REFERENCES

Axel Mithofer. 2022. Carnivorous plants and their biotic interactions. Journal of Plant Interactions, 17: 333–343.

Bochynek T, Burd M. 2024. Pollination efficiency and the pollen-ovule ratio. New Phytol 243(4): 1600–1609.

Ceccarelli S, Grando S. 2006. Decentralised-participatory plant breeding: An example of demand driven research. Euphytica 155: 349–360.

Chennai S et al. 2005. A predation behaviour model based on Game Theory. arXiv.

Christopher A Johnson et al. 2021. Coevolutionary transitions from antagonism to mutualism explained by the Co–Opted Antagonist Hypothesis. Natural Commun, 12: 2867.

Cucchi T, Arbuckle B. 2021. Animal domestication: from distant past to current development and issues. Anim Front 11(3): 6-9.

Daniela Pavlovic et al., 2014. Chlorophyll as a measure of plant health. Pesticidi i Fitomedicina, 29(1): 21-34.

Debarati Bhaduri et al. 2022. A review on effective soil health bio-indicators for ecosystem restoration and sustainability. Frontiers in Microbiology, 13: 1–25.

Fernie AR et al. 2006. Natural genetic variability for improving crop quality. Current Opinion in Plant Biology. 9(2): 196-202.

Filip J R Meysman et al. 2006. Bioturbation: a fresh look at Dawin's last idea. Trends Ecol Evol. 12: 688-695.

Ghosh S et al. 2020. Ecosystem Services of honey bees; regulating, provisioning and cultural functions. Journal of Apiculture. 35(2): 119.

Gleditsch, JM et al., 2022. Contemporizing island biogeography theory with anthropogenic drivers of species richness. Global Ecology and Biogeography, 32(2): 233-249.

Gruter C and Francis LW. 2011. Flower constancy in insect pollinators. Commun Integr Biol. 4(6): 633-636.

Heinz-Christian Frund et al., 2010. Earthworms as bioindicators of soil quality. In: Biology of Earthworms, pp 261-278.

Jenna M. Roper et al., 2021. Emerging technologies for monitoring plant health in vivo. ACS, Omega, 6(8): 5101-5107.

Kesavan PC, Swaminathan MS. 2020. Bio-village as a model of bioeconomy: In Current Developments in Biotechnology and Bioengineering. 7: 453-474.

Mariyappillai A et al. 2019. New agronomic ideas of eco-friendly pest management by insectivorous and murderous plants. https://ssrn.com/abstract=3500787

Mohanty AK et al. 2024. Evaluating the carnivorous efficacy of *Utricularia aurea* (Lamiales: Lentibulariaceae) on the larval stages of *Anopheles stephensi*, *Culex quinquefasciatus*, and *Aedes aegypti* (Diptera: Culicidae). J Med Entomol. 61(3): 719-725.

Morris CE et al. 2014. Bioprecipitation: A feedback cycle linking earth history, ecosystem dynamics and land use through biological ice nucleators in the atmosphere. Global Change Biology 20(2): 341-351.

Parcell J, Cain W. 2014. Ranking speciality crop profitability: Iterative stochastic dominance. Annual meeting, July 27-29, 2014, Minneapolis, Minnedota 170773, Agricultural and Applied Economics Association.

Powney GD et al. 2019. Widespread losses of pollinating insects in Britain. Nature Communications 10: 1018.

Purugganan MD. 2019. Evolutionary Insights into the nature of plant domestication. Curr Biol. 29(14): 705-714.

Sashika D et al, 2023. Fungi as environmental bioindicators. Science of the Total Environment 892: 164583.

Somerville V, et al. 2024. Genomic and phenotypic imprints of microbial domestication on cheese starter cultures. Nature Communications 15: 8642.

Thomson, J. D. 2001. Using pollination deficits to infer pollinator declines: Can theory guide us? Conservation Ecology 5(1): 6.

Weigelt P et al. 2019. GIFT - a global inventory of floras and traits for macroecology and biogeography. Journal of Biogeography 47(1): 16-43.

Wiranatha AS et al. 2024. Developing sustainable eco-agritourism based on resource based Theory in Klungkung Regency, Bali. International Journal of Tourism and Hotel Management. 6(1): 77-87.

Zhang J et al. 2023. De novo domestication: retrace the history of agriculture to design future crops. Current Opinion in Biotechnology 81: 102946.

Zulfiqar Ali et al. 2022. Wetting mechanism and morphological adaptation; leaf rolling enhancing atmospheric water acquisition in wheat crop-a review. Environ Sci Pollut Res. 29: 30967-30985.